Black Regulars in the War of 1812

Also by Eric Eugene Johnson

American Prisoners of War Held at Quebec During the War of 1812:
8 June 1813 – 11 December 1814

Ohio and the War of 1812:
A Collection of Lists, Musters and Essays

Ohio's Regulars in the War of 1812

Ohio's Black Soldiers Who Served in the Civil War

American Prisoners of War held in Montreal and Quebec
During the War of 1812

American Prisoners of War Paroled at Dartmouth, Halifax,
Jamaica and Odiham during the War of 1812

Black Regulars in the War of 1812

Eric Eugene Johnson

Society of the War of 1812
in the
State of Ohio

HERITAGE BOOKS
2015

HERITAGE BOOKS
AN IMPRINT OF HERITAGE BOOKS, INC.

Books, CDs, and more—Worldwide

For our listing of thousands of titles see our website
at
www.HeritageBooks.com

Published 2015 by
HERITAGE BOOKS, INC.
Publishing Division
5810 Ruatan Street
Berwyn Heights, Md. 20740

Heritage Books by the Society of the War of 1812 in the State of Ohio:

Translated by Harrison Scott Baker

American Prisoners of War Held at Bermuda, Cape of Good Hope and Jamaica During the War of 1812
American Prisoners of War Held at Barbados, Newfoundland and New Providence During the War of 1812
American Prisoners of War Held at Halifax, During the War of 1812, Volume I and II

Translated by Eric Eugene Johnson

American Prisoners of War Held in Montreal and Quebec during the War of 1812
American Prisoners of War Held at Quebec During the War of 1812, 8 June 1813–11 December 1814
Black Regulars in the War of 1812
Ohio and the War of 1812: A Collection of Lists, Musters and Essays
Ohio's Regulars in the War of 1812

International Standard Book Numbers
Paperbound: 978-0-7884-5610-7
Clothbound: 978-0-7884-6074-6

The Contents

The Introduction

The Black soldier was a rarity between the Revolutionary War and the Civil War. It is said that approximately 5,000 free Blacks and slaves served in the Continental Army during the Revolution. The Blacks proved their worth, many fought with distinction, and many died for their country.

During the Civil War, the U.S. Congress gave President Lincoln the authority to enlist Blacks as soldiers under the Militia Act of 1862.[1] 209,145 Black men would serve in the United States Colored Troops and in the United States Navy during this conflict. The Colored Troops would represent one-tenth of the total Union Army's strength during the war, while 40% of the Colored Troops would become casualties, either from the bullet or by disease.

Between these two wars, the Black man would be delegated to serve only as a company cook or a White officer's servant. The U.S. Navy would enlist Blacks on equal footing with the common sailors, however, the petty officer ranks and the officer ranks would still be off-limits to Blacks.

Cracks in this restriction appeared for a short time during the War of 1812. Without proper rules and regulations governing this issue, some commanders in the army did recruit Blacks as soldiers, although the total number of identified men is around 400. Louisiana did permit Blacks to serve in its state militia, and approximately, 800 Black men fought at the Battle of New Orleans in 1815.

The army did record, to some extent, on its enlistment rosters whether the soldier was White or Black. The recorded physical description of a soldier also indicated race, but this was not always accurate. The *Register of Enlistments in the U.S. Army, 1798-1914*[2] is an extract of personnel information from the land forces of the United States obtained from documents generated by the U.S. Army. These documents include enlistment records, recruiting reports, discharge records, commissioning records, muster rosters, inspection reports, morning reports, court-martial records, etc.

Prior to the War of 1812 there were Blacks serving as cooks and servants in the army. Black slaves serving as servants usually didn't have surnames, and nicknames were used as their given names, that is, Bob or Bill. In the comments section of the enlisted rosters, it may indicate that Bill was a Negro boy who was a servant for a particular officer while Bob was a Colored cook for a particular regiment.

The records generated during the War of 1812 have great genealogical value because the army recorded the age and place of birth, date and place of enlistment, physical descriptions, and the comment section highlighted the career of a soldier. Two problems exist: one, not all of the information is recorded for each soldier, and two, it is hard to spot a Mulatto (unless identified in a record), especially a very light skin Mulatto.

In the name column of the enlistment record there may be the words "negro," "black," "colored," or "mulatto" under the person's name. Likewise, these words may replace "private" in the rank column. "Colored" is often abbreviated as "Col'd" and Black as "Blk." The comments section also used these words but usually within sentences. Other terms that were used are "Negro boy," "Colored man," and "Blackman."

The army also captured the physical descriptions of a soldier by denoting the color of the eyes, the hair, and the complexion in the *Register of Enlistments*. During the 1800's, the term 'colored' denoted any one who wasn't white (Caucasian). In the records of the United States Colored Troops during the Civil War you will find Hawaiians, Mexicans and Filipinos, among other ethnic groups, serving with the Blacks.

---

[1] Sander, George P., **The Statues at Large, Treaties and Proclamations of the United States of America**, Volume XII, (Little, Brown and Company: Boston, MA 1863), pp. 597-600, Thirty-seventh Congress, session II, chapter 201, *An act to amend the Act calling forth the Militia to execute the Laws of the Union, suppress Insurrections, and repel Invasions, approved February twenty-eight, seventeen hundred and ninety-five, and the Acts amendatory thereof, and for other Purposes*, section 12 and 13.

[2] **Register of Enlistments in the U.S. Army, 1798-1914**; (National Archives Microfilm Publication M233, 81 rolls); Records of the Adjutant General's Office, 1780's-1917, Record Group 94; National Archives, Washington, D.C.

The color of the eyes will not denote race since all races had black eyes including Caucasians. Caucasians will also have red, blue, green, hazel, gray, and other colors for eyes. People of mixed races can also have eye color variants. Most races will have either a dark brown or black hair while Caucasians will have many different hair colors. In the army records, 'curly' will also denote a Colored person. The U.S. Navy enlistment records in the Civil War uses "wool" to identify Black sailors hair color.

The recording on the complexion of a soldiers is the best indicator of race. The color black is rarely used for a Caucasian. The terms used for a Caucasian are light, dark, fair, ruddy, sallow, freckled, flesh, red, sandy or florid. Other terms do exist. The terms used to describe Blacks or Mulattoes are black, brown, chestnut, or yellow. The last three terms usually denotes a Mulatto. The term 'yellow' is extremely interesting. In the 1800's, you were either white or Colored. "Yellow" was used to denote a mixed race person who wasn't fully white or fully Black. It did not mean that the person was Asian.

Using the three physical characteristics of a soldier may indicate either a Black or Mulatto soldier. These are:

Eyes	Hair	Complexion
Black	Black	Black
Black	Curly	Black
Black	Black	Yellow
Black	Curly	Yellow
Brown	Brown	Brown

Many Irish, Germans and men from the Mediterranean area of Europe are listed as having black or brown complexion along with black eyes and hair or brown eyes and hair. The place of birth is important to identify these men as Caucasian and not as Colored. However, Portugal and Spain were once slave nations and men identified with black or brown complexion could very well had been freed Blacks.

During the war, the British navy blockaded the coastline of the United States and bottled up most of our naval ships, merchant ships and privateers. The majority of Blacks recruited for the army were unemployed sailors who were sitting idle on the docks of Boston and Philadelphia. Most of these men were foreigners and many were born in the West Indies. Unless these men are listed as "Colored" in the enlistment records, it would be hard to prove that they were Colored simply because many of these men returned to the sea and possibly to their native countries after the war. However, many of them may have eventually settled in the United States and eventually became citizens.

A few of the Black men received land bounties and pensions after the war. In order to receive a land bounty, you had to apply for this benefit and you had to have your discharge papers. Many Blacks did not apply for a land bounty simply because they could not read or write, and others may have lost their discharge papers. Black men who enlisted with the 26th Regiment of U.S. Infantry in Philadelphia never joined the regiment in the field and they were not entitled to land bounties. Army requirements stated that all recruits had to join their regiments in the field in order to qualify for a land bounty. The war ended before these men were sent north from Philadelphia to join their regiment in New York. Some men did receive a land bounty but most had their bounties rejected by the War Department in 1820. Other Black soldiers who joined other regiments in New England did receive land bounties because they were sent from the recruiting depots to their regiments.

John Stagg, in his article on the enlisted men in the United States Army during the War of 1812, estimated that there were between 280 and 370 Blacks in the army.[3] His analysis of the army's *Register of Enlistments* showed that only 0.6% of those men whose skin color was noted had 'black' complexions. There were some men who were identified as "colored men' and their complexion were listed as either yellow or brown. Stagg stated that only 4,063 men had their physical descriptions recorded in the *Register*

[3] Stagg, John C. A., *Enlisted Men in the United States Army, 1812-1815: A Preliminary Survey*, **The William and Mary Quarterly**, Third Series, Volume XLIII, Number 4, October 1986, pp. 615-645.

of Enlistments, which leaves the possibility that many more Blacks can be found who enlisted in the army.

The majority of regiments found to have Blacks in their ranks, had only one or two men. The 26th Regiment of U.S. Infantry had the most Black men but this is not a factor since these men never joined the regiment in the field and they never served as combat soldiers. Most Blacks in this regiment have been identified due to the excellent record keeping of Captain William Bezeau who recruited the Blacks for the 26th Infantry.

This leaves the U.S. Light Artillery, the Corps of Artillery, and the 11th, 30th and 31st Infantries which enlisted more that five Blacks as possible candidates to have additional Blacks found within their ranks. Many of these Blacks served in the same companies together which may be an indicator that other Blacks can be found.

Every free able-bodied white male citizen

When Congress passed the Militia Act of 1792,[4] it required that "every free able-bodied white male citizen" join his state militia. It is implied that non-whites could not participate in the militia but it left open the possibility that non-whites could join the U.S. Army. On 5 April 1799,[5] the War Department established the requirements for its recruiting service by stating, "natives of good character, are always to be preferred for soldiers. Foreigners of good reputation, for sobriety and honesty, may be enlisted; but Negroes, Mulattoes, or Indians are not to be enlisted."

This new ruling only limited Blacks from being soldiers but they could still serve as cooks and servants. In 1802[6] Congress helped the officers who hired servants, or used their personal slaves as servants, by permitting the army to supply one ration (meal) per day for these individuals. Officers had always been allowed to take a private from within the officer's command to serve as a servant. These privates serving in this capacity were usually called "waiters."

Congress in 1812, when passing legislation to provide for the army, increased support for officer's servants by stating, "that the officer's who shall not take waiters from the line of the army, shall receive the pay, clothing and subsistence allowed to a private soldier." [7] This was a big boost for Black servants, for the U.S. Army would now feed, pay and cloth these men on the same level as a private. Servants would now receive two rations per day, the same as provided to the privates.

Legislation in 1814 forbid officers from employing soldiers as servants and stated that the names of the servants must now appear on the muster rolls for each corps.[8] Servants were now full members of the U.S. Army. This legislation was not enforced and the names of most servants, white or black, continued to be left off most of the muster rolls.

Cracks in these restrictions appeared for a short time during the War of 1812. Congress passed *An Act for Completing the Existing Military Establishment*[9] on 24 December 1811 in which is stated that only "able bodied men" may be recruited in the army. No restrictions for race will appear in any military legislation passed during the War of 1812.

Without proper rules and regulations governing this issue of recruiting Black soldiers, some commanders in the army did recruit Blacks although the total number of men was around 400. Louisiana

---

[4] Peters, Richard, **The Public Statutes at Large of the United States of America**, Volume I, (Charles C. Little and James Brown: Boston, MA 1845), pp.271-274, Second Congress, 1st session, chapter 33, 8 May 1792, *An Act more effectually to provide for the National Defence by establishing an Uniform Militia throughout the United States*, section 1.

[5] **War Department, Rules and Regulations Respecting the Recruiting Service**, April 5, 1799, Article III – Natives of good character, are always to be preferred for soldiers. Foreigners of good reputation, for sobriety and honestly, may be enlisted; but Negroes, Mulattoes, or Indians are not to be enlisted.

[6] Peters, Richard, **The Public Statutes at Large of the United States of America**, Volume II, (Charles C. Little and James Brown: Boston, MA 1845), pp.132-137, Seventh Congress, 1st session, chapter 9, 16 March 1802, *An Act fixing the Military Peace Establishment of the United States*, section 5.

[7] Peters, Richard, **The Public Statutes at Large of the United States of America**, Volume II, (Charles C. Little and James Brown: Boston, MA 1845), pp.784-785, Twelfth Congress; 1st session, chapter 137, 6 Jul 1812, *An Act making further Provision for the Army of the United States, and for other purposes*, section 5.

[8] Peters, Richard, **The Public Statutes at Large of the United States of America**, Volume III, (Charles C. Little and James Brown: Boston, MA 1846), pp.113-116, Thirteenth Congress, 2nd session, chapter 37, 30 March 1814, *An Act for the better organizing, paying, and supplying the Army of the United States*, section 10.

[9] Peters, Richard, **The Public Statutes at Large of the United States of America**, Volume II, (Charles C. Little and James Brown: Boston, MA 1845), pp.669-680, Twelfth Congress, 1st session, chapter 10, 24 December 1811, *An Act for Completing the Existing Military Establishment*, section 2.

did permit free Blacks to serve in the state militia, and approximately 800 Black men fought in the Battle of New Orleans. The total number of Blacks possible serving in the army and the state militias was likely less than 1,500 men. There were maybe another 500 light-skinned Mulattoes in the rest of the military forces.

All Blacks were discharged in May of 1815 regardless of their enlistment obligations. Some of the comments in the *Register of Enlistments* state that Colonel Henry Atkinson of the 6th Regiment of U.S. Infantry issued the discharge order on behalf of the War Department. The author has found no general orders from the Department of War ordering Blacks to be dismissed from the service. Most likely, this was a general order issued by the Northern Army in New York.

The Adjutant General's Department of the Department of War did issue a general order on 18 February 1820 stating that "no negro or mulatto will be received as a recruit of the army."[10] This order does not direct officers from discharging Negroes or Mulattoes, only that they can no longer recruit these men. Regardless, this general order ended the Black man's participation as a soldier in the army until the Civil War.

Louisiana's Free Men of Color

Louisiana was the only state which permitted free Blacks to become members of the state militia. This is probably due to the fact that Louisiana was the only state not established under English laws but under French Napoleonic Codes. Louisiana's militia laws were 'grandfathered' so that the state did not have to abide by the Militia Act of 1792 restricting Blacks from serving in the state militia.

A common myth is that Andrew Jackson created the two Black militia battalions that served under him during the Battle of New Orleans. Both battalions had existed before the war. On 21 September 1814 while at his headquarters in Mobile, Mississippi Territory (now Alabama), Jackson issued a proclamation to the 'Free Colored Inhabitants of Louisiana' to join their state militia.[11] On 18 December 1814 while at New Orleans, Jackson issued a broadside calling for free Blacks to join this army. Both battalions were under strength and Jackson was able to bring these units up to full strength in time for the Battle of New Orleans. The two battalions were under the commands of Major Pierre Lacoste and Louis D'Aquin, and they consisted of Black officers and Black enlisted men.

Thomas Moebs in his book, *Black Soldiers – Black Sailors – Black Ink,* lists the men who served in these two Louisiana Free Black battalions.[12] He also lists some of the men who enlisted in the U.S. Army and in the U.S. Navy during the War of 1812.

The Black regiments of New York and Pennsylvania

Both New York and Pennsylvania (Philadelphia) raised Black regiments during the War of 1812, but these units were not military regiments. These regiments were construction battalions used to create military fortifications around New York City, south of Philadelphia, and at Sackets Harbor, NY.

The New York legislature passed *An Act to authorize the raising of two regiments of Men of Color* on 24 October 1814 to serve for three years or less.[13] Each regiment would have 1,080 men consisting of white officers with free and slave Colored men. The slaves would be entitled to be freed upon discharge.

---

[10] Drum, Richard C. (Brigadier General), **Adjutant General's Department**, Subject Index of the General Orders of the War Department from January 1, 1809 to December 31, 1860, page 122, 18 February 1820, general order, "no negro or mulatto will be received as a recruit of the army."

[11] Wilson, Joseph T., **The Black Phalanx: A History of the Negro Soldiers of the United States in the Wars of 1775-1812, 1861-1865**, (American Publishing Company: Hartford, CT, 1890), Chapter II, The War of 1812, pp. 72-80.

[12] Moebs, Thomas Truxtun, **Black Soldiers – Black Sailors – Black Ink: Research Guide of African-Americans in U.S. Military History, 1512-1900**, (Moebs Publishing Co.: Chesapeake Bay, VA 1994).

Congressman Henry Martindale of New York said in a speech delivered on the 22 January 1828 before the U.S. House of Representatives that "slaves or negroes who had been slaves were enlisted as soldiers in the war of the Revolution: and I myself saw a battalion of them, as fine martial looking men as I ever saw attached to the Northern army in the last war (1812), on its march from Plattsburg to Sacket's Harbor, where they did service for the country with credit to New York and honor to themselves." [14]

The Pennsylvania regiment was recruited by three Black Philadelphians for the defense of this city.[15] An excerpt from the *Missing Pages in American History* states "armed by the conduct of the foe at Washington and Baltimore, many of the other coast cities began to make preparation for their own protection. In Philadelphia, the Vigilance Committee solicited the aid of Absolem Jones, Richard Allen and James Forten, all Negroes. It was desired that these men should secure the services of members of their own race in erecting defences about the city. Through their efforts twenty-five hundred black men were gathered in the yard of the State House in August and marched to Grey's Ferry, on the west side of the Schuylkill River, where for two days they were employed in throwing up fortifications, for which they received a vote of thanks, tendered them by the city. Mr. Forten himself worked on these breastworks with twenty of his journeymen. A body of colored troops was organized there at this time and placed under an officer of the United States Army."

British Black regiments

Britain raised the Corps of Colonial Marines during the War of 1812 which included runaway American slaves from the Chesapeake Bay area and later along the Georgia coast. Two battalions of Black marines were created. One battalion of marines served in the Chesapeake Bay area while the other battalion served off the east coast of Florida. Upon discharge at war's end, the northern battalion of 700 men settled in Trinidad while the southern battalion of 300 men stayed in Florida and eventually became a part of the Black Seminoles.

The British also sent three Black West Indian Regiments to Louisiana before the Battle of New Orleans. These were the 1st, 2nd and 5th West India Regiments. In Canada, the 104th Regiment of Foot enlisted Black pioneers. Pioneers were soldiers who were armed with an ax and a shovel. They cleared a path for the regiment in woods or over rough terrain.

The Canadian militia also had a company of Blacks serving along the Niagara frontier.[16] Captain Robert Runchey commanded the Company of Coloured Men in the 1st Regiment of Lincoln Militia of Upper Canada (now Ontario). This company would first serve as infantrymen but would later be formed into the Corps of Artificers in early 1813.

American Service

This book highlights 398 known African Americans who served in the U.S. Army during the War of 1812. Many servants are not listed on the musters rolls of the various regiments as was required by army regulations. Many Mulattoes are hidden in the enlistment records because they are very light skinned and could pass as Caucasian. This leaves the possibility of identifying more Blacks in the future who enlisted

---

[13] Williams, George W., **History of the Negro Race in America from 1619 to 1880**, (G. P. Putnam's Sons: New York, 1883), volume II, chapters 2 and 3, pages 23-27, Negro Troops in the War of 1812 and Negroes in the Navy.

[14] Wilson, Joseph T., **The Black Phalanx: A History of the Negro Soldiers of the United States in the Wars of 1775-1812, 1861-1865**, (American Publishing Company: Hartford, CT, 1890), Chapter II, The War of 1812, pp. 72-80.

[15] Wilkes, Laura E., **Missing Pages in American History: Revealing the Service of Negroes in the Early Wars in the United States of America, 1641-1815**, (Press of R. L. Pendleton: Washington, DC, 1919), War of 1812, pages 65-66, Philadelphia.

[16] Irving, L. Homfray, **Officers of the British Forces in Canada during the War of 1812**, (Welland Tribune Printers, 1908), page 76, The Upper Canada Militia, Captain Runchey's Company of Coloured Men.

in the army during this war. Some of these men will only be identified though family histories and traditions.

Very few of these men ever saw any type of combat. Of the 396 men, 258 were recruits and they never left Philadelphia. Of the remaining 140 men, eighteen were officer's servants which leaves 121 men. Of the fifty-four men who served in the U.S. Light Artillery, the majority of these men served at Fort Independence in Boston harbor. Apparently, these men were used to free white soldiers for combat. This appears to hold true for the men of the 10th Regiment of U.S. Infantry who were stationed at Fort Washington, MD.

Three Blacks died during their enlistment, not from wounds received during battle but from disease. Two men became prisoners of war. One of these men was wounded. Another man was wounded during the Battle of Chippewa and a final man was wounded during one of the battles at Fort Erie. Both of these battles occurred in Ontario, Canada. One Black soldier served with the U.S. naval squadron on Lake Champlain on temporary duty.

Of the seventy-three land bounties issued to Black men for service during the war, sixty-one of these were for men in the 26th Regiment of U.S. Infantry. These men were recruits and were not entitled to land bounties according to army regulations.[17] Ten of these land bounties were rejected and five were re-issued. The majority of these land bounties were issued after 1850 when the militia was able to apply for land bounties due to War of 1812 service. Three men received a pension for their service. Jordan B. Noble received both a land bounty and a pension due to his service during the Mexican War.

Some commanders were not comfortable having Blacks serving in the U.S. Army. Major General George Izard was in charge of the Northern Army in New York during the last year of the war. He tried to transfer all of the Blacks under his command into a Pioneer Corps, which would do the manual labor for his army. In his letter to the Secretary of War on 3 July 1814,[18] he stated, "there was, some years ago, a regulation of our service, prohibiting the enlistment of negroes and people of colour. I have not heard of its being enforced. Among the New England recruits there have lately been brought hither a number of these people, to the great annoyance of the officers and soldiers here. The latter object to doing duty with them. The Inspector General is now organizing them as a sort of pioneer corps. Shall they be retained and mustered in that capacity?" Apparently this never happened! None of the comments included in each of the men's service records indicates that any of the Colored men were transferred a Pioneer Corps.

---

[17] Drum, Richard C. (Brigadier General), **Adjutant General's Department**, Subject Index of the General Orders of the War Department from January 1, 1809 to December 31, 1860, page 26, 4 March 1815, general order, Recruits enlisted for the war, and discharged without joining any regiments or corps, will not receive the retained bounty.

[18] **Official Correspondence with the Department of War relative to the Military Operations of the American Army under the Command of Major General Izard of the Northern Frontier of the United States in the Years 1814 and 1815**, (William Fry, Printer: Philadelphia, 1816), pp. 45-46, Letter from Major General Izard to the Secretary at War, 3 July 1814.

The Regiments

10th Regiment of U.S. Infantry

11 January 1812 – 17 May 1815

The 10th Regiment of U.S. Infantry was organized under the Acts of 11 January 1812 and 26 June 1812. Under the Act of 3 March 1815, the regiment was consolidated with the 8th, 36th and 38th Regiments of Infantry to form the new 6th Regiment of Infantry on 17 May 1815.

The regiment was raised primarily in North Carolina and enlistments were for five years, eighteen months or 'During the War.' The recruiting headquarters was located in Wilkesboro, North Carolina. The regiment participated in the Battle of Chateaugay, Upper Canada, on 26 October 1813 and in the Battle of La Cole Mill, Lower Canada, on 30 March 1814.

11th Regiment of U.S. Infantry

11 January 1812 – 17 May 1815

The 11th Regiment of U.S. Infantry was organized under the Acts of 11 January 1812 and 26 June 1812. Under the Act of 3 March 1815, the regiment was consolidated with the 25th, 27th, 29th and 37th Regiments of Infantry to form the 6th Regiment of Infantry on 17 May 1815.

The regiment was raised in Vermont, New Hampshire and Connecticut. Recruiting headquarters were established in Rutland and Bennington, Vermont, with enlistments or five years, eighteen months or 'During the War.' The 11th Infantry saw action in the following battles: Crystler's Fields, La Cole Mill, Fort Erie, Chippewa, and Lundy's Lane. All of these battles were in Upper Canada.

26th Regiment of U.S. Infantry

30 March 1814 to 17 May 1815

The 26th Regiment of U.S. Infantry was organized from the U.S. Voluntary Corps under the provision of section twenty-one of the Act of 30 March 1814 as the 48th Regiment of U.S. Infantry. The regiment was re-designated as the 26th Regiment of U.S. Infantry on 12 May 1814 under section 13 of the Act of 30 March 1814. The regiment was consolidated on 17 May 1815 under the Act of 3 March 1815 with the Regiment of Light Artillery.

The regiment was assigned to be raised in Vermont but it maintained recruiting stations in New England, New York, Philadelphia and Baltimore. The recruiting headquarters was located in Burlington, Vermont. Enlistments were for five years or eighteen months. Armed with rifles, the 26th Infantry participated in the Battles of Fort Erie in August and September of 1814.

Captain William Bezeau was the recruiting officer for the 26th Infantry at Philadelphia. He recruited 258 Blacks between August 1814 and June 1815. The captain also recruited Caucasians which he kept in a separate detachment. His detachments were stationed at Lareretto Barracks, south of Philadelphia in Tinicum Township, Delaware County. The barracks were a part of the Lareretto Quarantine Station which was a hospital treating epidemics both locally and those brought by ships.

Many of the Blacks were detached to work in the station's hospital which was probably used as a military hospital during the war. None of the Black recruits were sent north to join the regiment in New York. At the war's end, those Blacks who enlisted "during the war" were discharged between 20 and 31 May 1815 while the men who enlisted for five years were released between 1 May and 1 June 1815.

The names of most of the Blacks can be found on one of two descriptive rolls maintained for this detachment. The first descriptive roll was dated 1 April 1815 while the other is dated 25 May 1815. The Caucasians were listed on separate descriptive rolls.

30th Regiment of U.S. Infantry

29 January 1813 to 17 May 1815

The 30th Regiment of U.S. Infantry was organized under the Act of 29 January 1813. Under the Act of 3 March 1815 the regiment was consolidated with the Regiment of Light Artillery on 17 May 1815.

The recruiting headquarters was located at Burlington, Vermont. Originally a one-year regiment with one-year enlistments, the 30th Infantry was converted to a standard regiment with five years or eighteen months enlistments on 28 January 1814. The regiment saw action at La Cole Mill, Lower Canada, and later at Plattsburg, NY, between 6 and11 September 1814.

31st Regiment of U.S. Infantry

29 January 1813 to 17 May 1815

The 31st Regiment of U.S. Infantry was organized under the Act of 29 January 1813. Under the Act of 3 March 1815 the regiment was consolidated with the Regiment of Light Artillery on 17 May 1815.

The recruiting headquarters was located at Woodstock, Vermont. Originally a one-year regiment with one-year enlistments, the 31st Infantry was converted to a standard regiment with five years or months months enlistments on 28 January 1814. The regiment saw action at Chateaugay and La Cole Mill in Lower Canada, and at Plattsburg, NY.

Regiment of U.S. Light Artillery

12 April 1808 to 1 June 1821

The Regiment of U.S. Light Artillery was organized under the Act of 12 April 1808 to consist of ten companies. Captain George Peter's company was the only company of this regiment, which prior to the War of 1812 equipped as field artillery. It was mounted as such in 1808 and dismounted in 1809. When the War of 1812 broke out some of the companies were equipped, only for a short time served, as horse artillery and others as field artillery. The Light Artillery served primarily as an elite infantry regiment.

On 17 May 1815, under the Act of 3 March 1815, the 15th, 26th, 30th, 31st, 34th and 45th Regiments of Infantry were consolidated into this regiment. On 1 June 1821 under the Act of 2 March 1821, the regiment was disbanded and its companies were distributed among the 1st, 2nd, 3rd, and 4th Regiments of Artillery.

The regiment was raised in the northeastern part of the United States and it had its recruiting headquarters located at Dedham, Massachusetts. Recruitments were for five years or during the war. The Light Artillery participated in the battles at Queenston Heights, Fort George, Stoney Creek and Beaver Dams in Upper Canada, plus French Creek in New York, Crystler's Fields, La Cole, and at Fort Oswego, New York.

The regiment listed the second largest number of Blacks of any U.S. regiment, fifty-four men. The majority of these men were enlisted in Boston by William Campbell and Samuel Washburn. The men served in companies commanded by John Bell, John McIntosh and George Morris. Nearly all of the men served at Fort Independence in Boston harbor. All of the men were discharged on 31 May 1815 regardless of their enlistment enrollments.

Corps of Articifers

23 April 1812 to 3 March 1815

The Corps of Artificers was organized under the Act of 23 April 1812 and the unit was assigned to the Quartermaster's General Department. The Corps was discharged under the Act of 3 March 1815.

Alexander Paris was the superintendent of this company size unit. The Corps consisted of masons, carpenters, blacksmiths, boat builders, armorers, saddle and harness markers, plus laborers. The unit was designed to assist the army in constructing any type of building, fortifications, or boats. Eleven Blacks were assigned as laborers in this Corps.[19]

[19] Clark, Byron N., **A List of Pensioners of the War of 1812**, (Research Publications Company: Burlington and Boston, 1904), *Payroll of a Company of Artificers, Commanded by Alexander Paris, Superintendent of the Corps of Artificers of the United States, for the months of May and June 1814*, pp. 129-131.

The Scorecard

This chapter will explain the data fields, abbreviations, terms and phrases used in creating the 'service record' of each individual listed in the following two chapters.

Data Field	**Explanation**
Rank	The highest known military rank is listed for each soldier.
Regiment	The name of the regiment which the soldier was a member.
Other regiment(s)	If a soldier was transferred to another regiment or regiments, the name of the additional regiments are listed in this field.
Company	The company commander's name
Age	Age of the soldier at the time of his enlistment
Height	Height of the soldier in feet and inches.
Birth Place	The enlistment rosters list the birth place of a soldier by state or country, county and city.
Trade	Civilian trade of a soldier at the time of enlistment.
Enlistment date	Date that an enlisted man entered service.
Enlistment Place	The place of enlistment by state, county and city.
Enlistment Period	There were four enlistment periods which a soldier could select. All soldiers could choose to re-enlist once their initial enlistment period ended. Those soldiers who enlisted "during the war" were discharged at the end of the War of 1812. 18 Mos = 18 months 1 Yr = 1 year 5 Yrs = 5 years War = "during the war"
By whom	The recruiting officer's name.
Discharged	The date of discharge and location along with other comments.
Died or Killed	Date of death and known location while serving in the army with additional comments.
Pension	I-9999 – Invalid IC-9999 – Invalid's Certificate IF-9999 – Invalid's File IO-9999 – Invalid's Original MC-9999 – Minor's Certificate MO-9999 – Minor's Original SC-9999 – Survivor's Certificate SF-9999 – Survivor's File SO-9999 – Survivor's Original

	WC-9999 – Widow's Certificate WF-9999 – Widow's File WO-9999 – Widow's Original
Bounty Number	BLW 123456-160-12 BLW = Bounty land warrant number -160- or -320- = number of acres issued for the warrant -12 or -14 or -42 or -50 or -55 = Years of the Land Bounty Acts (Arkansas), (Illinois), (Missouri), or other = State in which the warrant was assigned
Comments	Any additional comments for a soldier.
Bounty or land bounty	Regular soldiers were entitled to either 160 or 320 aces of free land for their service during the War of 1812. The heirs of the soldiers, who were killed or who had died in service, were also entitled to bounties.
Descriptive Roll	Each company maintained a Descriptive Roll listing every man in the company, his age, height, weight, color of eyes and complexion, where born, where enlisted, the name of the officer who enlisted them, enlistment date, length of enlistment, and bonus amounts that were paid and what was due. These reports could also include a list of clothing and other supplies that were issued to the soldiers.
Double Bounty	Soldiers who enlisted or re-enlisted on or after 1 February 1814 were entitled to 320 acres of bounty land instead of 160 acres.
Pioneer	A pioneer was a soldier who was delegated to clear a path for a regiment, normally in a wooded area. These soldiers wore a leather apron and used axes and shovels to clear a road.
Recruit	A private still at a recruiting headquarters who had not been assigned to a company and who had not been assigned to his regiment in the field.
Servant	A civilian or slave who served the needs of an officer. This person maintained the household (tent, room or apartment) of an officer and might also perform washing and cooking duties.
Service Record	A service record was any document found in the *Register of Enlistments* which highlighted the service of a soldier.
Surgeon's Certificate of Disability	This was a medical discharge given to a soldier by a regimental surgeon.
Waiter	A waiter was a soldier taken from an officer's command who performed the duties of a servant.

Standard U.S. Postal Service abbreviations are used for U.S. states plus

LC = Lower Canada (now Quebec)
UC = Upper Canada (now Ontario)
NS = Nova Scotia

Heirs obtained half pay for five years in lieu of military bounty land

Early in the war, the heirs of soldiers who had died (or were killed) during the war and who had enlisted for one-year or 18-months, could elect to receive the soldiers' half-month pay for five years or receive 160 acres of **land**. This was changed early in the war so that the heirs only received the half pay. The heirs of soldiers who enlisted for five years or during the war always received either 160 or 320 acres of land. 320 acres of land was issued to soldiers who enlisted or re-enlisted after 1 February 1814.

Land bounty to "name of heirs" heirs at law of "name of soldier"

This phase lists the name(s) of the heir or heirs by law who received the land bounty of a deceased soldier.

- In memory of those who did not return -

The Honored Dead

David Davis
Aaron Forman
James Gomaus
David James
Henry Jones
John Moore
Warner Wallace
Philip Winsell

- Those who die in service to the United States should not be forgotten –

The Soldiers

-----, Alfred - Waiter - U.S. Rifles - Company: Benjamin Desha - Enlistment date: 1 Jul 1814 - Comments: Listed as a "Mulatto" in his service record; waiter to First Lieutenant John Heddelson.

-----, Andrey - Servant - General Staff - Company: James Hamilton - Comments: Private servant to Captain James Hamilton, major assistant inspector general; listed on the payroll report of 31 Jul 1814 (*A List of Pensioners of the War of 1812*).

-----, Bill - Waiter - U.S. Quartermaster Department - Detachment: Nehemiah Baden - Comments: Listed as a Negro boy in his service record; private waiter to Captain Gustavus Harrison, assistant deputy quartermaster general, Washington, DC.

-----, Bob - Waiter - U.S. Quartermaster Department - Detachment: Nehemiah Baden - Comments: Listed as a Negro boy in his service record; private waiter to Major Marstella, Washington, DC (probably Ferdinand Marsteller, assistant deputy quartermaster general).

-----, David - Servant - General Staff - Company: Alexander Macomb - Comments: Private servant to Major General Alexander Macomb; listed on the payroll report of 31 Jul 1814 (*A List of Pensioners of the War of 1812*).

-----, Edward - Servant - 2nd U.S. Light Dragoons - Company: George Haig - Comments: Listed on the Muster Roll of 30 Jun 1814 (*A List of Pensioners of the War of 1812*).

-----, Henry - Servant - 2nd U.S. Light Dragoons - Company: George Haig - Comments: Listed on the Muster Roll of 30 Jun 1814 (*A List of Pensioners of the War of 1812*).

-----, Isaac - Servant - 2nd U.S. Light Dragoons - Company: George Haig - Comments: Listed on the Muster Roll of 30 Jun 1814 (*A List of Pensioners of the War of 1812*).

-----, Lewis - Waiter - U.S. Ordnance Department - Detachment: Nehemiah Baden - Comments: Negro boy at Washington, DC, in First Lieutenant Nehemiah Baden's Detachment serving as a private waiter.

-----, Moses - Servant - 2nd U.S. Light Dragoons - Company: George Haig - Comments: Listed on the Muster Roll of 30 Jun 1814 (*A List of Pensioners of the War of 1812*).

-----, Nace - Waiter - U.S. Ordnance Department - Detachment: Nehemiah Baden - Comments: Listed as a Negro boy in his service record; private waiter to First Lieutenant Nehemiah Baden, Assistant Deputy Commissary of Ordnance at Washington, DC.

-----, Reuben - Waiter - U.S. Ordnance Department - Detachment: Nehemiah Baden - Comments: Listed as a Negro boy in his service record; private waiter to Capt Morton (probably Captain John Morton, Deputy Commissary of Ordnance).

-----, Robert - Waiter - Unknown - Enlistment date: 8 Jun 1814 - Comments: Listed as a Negro boy in his service record; private waiter in Washington, DC.

-----, Shadrach - Waiter - U.S. Ordnance Department - Detachment: Nehemiah Baden - Comments: Listed as a Negro boy in his service record; private waiter to First Lieutenant Nehemiah Baden,

Assistant Deputy Commissary of Ordnance at Washington, DC.

Africanus, Major - Servant - 24th U.S. Infantry - Company: Robert Desha - Comments: Private servant of First Lieutenant Wyly Martin (probably Colored due to his surname).

Alarkin – see Manuel A. Larkin

Aldridge, John - Recruit - 26th U.S. Infantry - Recruiting Detachment: William Bezeau - Age: 22 - Height: 5' 7" - Eyes: Black - Hair: Curly - Complexion: Chestnut - Born: Santo Domingo (Haiti) - Trade: Laborer - Enlistment date: 9 Jan 1815 - Enlistment place: Philadelphia - Enlistment period: War - Enlisted by whom: William Bezeau - Discharge date: 20 Mar 1815 - Bounty: BLW 804-320-12 (Missouri) - Comments: Listed on the Descriptive Roll of Colored Men, 1 Apr 1815; land bounty to Betty Aldridge, sister and only heir at law of John Aldridge; he died before 21 Jun 1819.

Alexander, John - Recruit - 26th U.S. Infantry - Recruiting Detachment: William Bezeau - Age: 23 - Height: 6' - Eyes: Black - Hair: Curly - Complexion: Black or Chestnut - Born: Pennsylvania - Trade: Laborer - Enlistment date: 12 Nov 1812 - Enlistment place: Philadelphia - Enlistment period: 5 Yrs - Enlisted by whom: William Bezeau - Discharge date: 1 May 1815 - Comments: Listed as a "Negro man" in his service record; discharged at Philadelphia, PA.

Alfred, John - Private - 30th U.S. Infantry - Company: Daniel Farrington - Age: 20 - Height: 5' 11" - Eyes: Black - Hair: Black - Complexion: Black - Born: VT, Franklin County, Shelden - Trade: Farmer - Enlistment date: 24 Mar 1814 - Enlistment place: NY, Plattsburg or VT, Swanton - Enlistment period: War - Enlisted by whom: Simeon Wright - Discharge date: 17 Jun 1815 - Bounty: BLW 7837-160-42 (Illinois) - Comments: Possible Black, not verified; discharged at Plattsburgh or Champlain Station, NY.

Allen, George - Recruit - 26th U.S. Infantry - Recruiting Detachment: William Bezeau - Age: 16 - Height: 5' - Eyes: Black - Hair: Black - Complexion: Black - Born: Pennsylvania - Trade: Farmer - Enlistment date: 9 Oct 1814 - Enlistment place: Philadelphia - Enlistment period: 5 Yrs - Enlisted by whom: William Bezeau - Comments: Listed as a "Colored man" in his service record; deserted on 18 Oct 1814.

Allen, James - Recruit - 26th U.S. Infantry - Recruiting Detachment: William Bezeau - Age: 22 - Height: 5' 8" - Eyes: Black - Hair: Black or Brown - Complexion: Brown - Born: PA, Philadelphia - Trade: Laborer - Enlistment date: 21 Sep 1814 - Enlistment place: Philadelphia - Enlistment period: 5 Yrs - Enlisted by whom: William Bezeau - Discharge date: 16 May 1815 - Comments: Listed as a "Negro man" in his service record.

Allen, Peter - Recruit - 26th U.S. Infantry - Recruiting Detachment: William Bezeau - Age: 23 - Height: 5' 9" - Eyes: Black - Hair: Curly - Complexion: Chestnut - Born: Pennsylvania - Trade: Farmer - Enlistment date: 11 Jan 1815 - Enlistment place: Philadelphia - Enlistment period: War - Enlisted by whom: William Bezeau - Bounty: BLW 830-320-14 Cancelled - Comments: Listed on the Descriptive Roll of Colored Men, 1 Apr 1815; bounty cancelled by an order from the War Department on 19 May 1820.

Allen, William - Recruit - 26th U.S. Infantry - Recruiting Detachment: William Bezeau - Age: 21 - Height: 5' 9" - Eyes: Black - Hair: Black - Complexion: Black - Born: Delaware - Trade: Laborer - Enlistment date: 5 Nov 1814 - Enlistment place: Philadelphia - Enlistment period: War - Enlisted by whom: William Bezeau - Discharge date: 20 Mar 1815 - Bounty: BLW 7212-160-12 (Illinois) - Comments: Listed as a "Colored man" in his service record; discharged at Philadelphia, PA.

Alley, William - Recruit - 26th U.S. Infantry - Recruiting Detachment: William Bezeau - Age: 28 - Height: 5' 10" - Eyes: Black - Hair: Curly - Complexion: Brown - Born: Georgia - Trade: Farmer - Enlistment date: 6 Jan 1815 - Enlistment place: Philadelphia - Enlistment period: War - Enlisted by whom: William Bezeau - Comments: Listed on the Descriptive Roll of Colored Men, 1 Apr 1815.

Anderson, Joseph - Recruit - 26th U.S. Infantry - Recruiting Detachment: William Bezeau - Age: 21 - Height: 5' 4" - Eyes: Black - Hair: Curly - Complexion: Black - Born: Louisiana - Trade: Seaman - Enlistment date: 12 Jan 1815 - Enlistment place: Philadelphia - Enlistment period: War - Enlisted by whom: William Bezeau - Comments: Listed on the Descriptive Roll of Colored Men, 1 Apr 1815.

Anderson, Samuel - Recruit - 26th U.S. Infantry - Recruiting Detachment: William Bezeau - Age: 20 - Height: 5' 5" - Eyes: Black - Hair: Black - Complexion: Black - Born: Virginia - Trade: Laborer - Enlistment date: 5 Sep 1814 - Enlistment place: Philadelphia - Enlistment period: 5 Yrs - Enlisted by whom: William Bezeau - Discharge date: 17 May 1815 - Comments: Listed as a "Colored man" in his service record; discharged at Philadelphia, PA.

Attire, Peter - Recruit - 26th U.S. Infantry - Recruiting Detachment: William Bezeau - Age: 19 - Height: 5' 6" - Eyes: Black - Hair: Curly - Complexion: Black - Born: Maryland - Trade: Farmer - Enlistment date: 2 Jan 1815 - Enlistment place: Philadelphia - Enlistment period: War - Enlisted by whom: William Bezeau - Comments: Listed on the Descriptive Roll of Colored Men, 1 Apr 1815; deserted on 18 Jan 1815.

Baldwin, Cyrus (Bladen) - Recruit - 26th U.S. Infantry - Recruiting Detachment: William Bezeau - Age: 24 - Height: 5' 9" - Eyes: Black - Hair: Black - Complexion: Black - Born: Pennsylvania - Trade: Laborer - Enlistment date: 10 Oct 1814 - Enlistment place: Philadelphia - Enlistment period: War - Enlisted by whom: William Bezeau - Discharge date: 20 Mar 1815 - Bounty: BLW 7183-160-12 (Illinois) - Comments: Listed as a "Colored man" in his service record; discharged at Philadelphia, PA.

Baptist, John - Laborer - Corps of Artificers - Company: Alexander Paris - Comments: Listed as "Black" in *A List of Pensioners of the War of 1812*.

Baptiste, John (or John Baptiste Delaurier) - Recruit - 26th U.S. Infantry - Recruiting Detachment: William Bezeau - Age: 28 - Height: 5' 2" - Eyes: Black - Hair: Curly - Complexion: Black - Born: Africa - Trade: Hatter - Enlistment date: 1 Jan 1815 - Enlistment place: Philadelphia - Enlistment period: War - Enlisted by whom: William Bezeau - Discharge date: 20 Mar 1815 - Bounty: BLW 534-320-14 (Illinois) - Comments: Listed on the Descriptive Roll of Colored Men, 1 Apr 1815; discharged at Philadelphia, PA.

Barager, Francis (Barragher) - Recruit - 26th U.S. Infantry - Recruiting Detachment: William Bezeau - Age: 39 - Height: 5' 6" - Eyes: Hazel - Hair: Black - Complexion: Yellow - Born: New Jersey - Trade: Laborer - Enlistment date: 10 Dec 1814 - Enlistment place: Philadelphia - Enlistment period: War - Enlisted by whom: William Bezeau - Discharge date: 20 Mar 1815 - Comments: Listed as "Colored" in his service record; discharged at Philadelphia, PA.

Barber, James - Private - U.S. Light Artillery - Company: George Morris - Age: 40 - Height: 5' 4 1/2" - Eyes: Black - Hair: Black - Complexion: Black - Born: NY, New York - Trade: Laborer - Enlistment date: 7 Oct 1814 - Enlistment place: MA, Boston - Enlistment period: War - Enlisted by whom: William Campbell - Discharge date: 31 Mar 1815 - Comments: Possible Black, not verified.

Barkess – see Jacob Burkess

Barkley, Benjamin (Barclay) - Recruit - 26th U.S. Infantry - Recruiting Detachment: William Bezeau - Age: 23 - Height: 5' 7 1/2" - Eyes: Black - Hair: Black - Complexion: Yellow - Born: Delaware - Trade: Laborer - Enlistment date: 9 Oct 1814 - Enlistment place: Philadelphia - Enlistment period: 5 Yrs - Enlisted by whom: William Bezeau - Comments: Listed as "Col'd" in his service record.

Barnwell, George - Private - 23rd U.S. Infantry - Company: Azariah Odell - Other regiment: 2nd U.S. Infantry - Age: 20 - Height: 5' 7" - Eyes: Black - Hair: Black - Complexion: Black - Born: LA, New Orleans - Trade: Sailor - Enlistment date: 6 Sep 1814 - Enlistment place: NY, Utica - Enlistment period: War - Enlisted by whom: Alphonso Wetmore - Comments: Possible Black, not verified; transferred to Captain Peter Van Buren's Company, 2nd U.S. Infantry on 30 Apr 1815; deserted from Sackets Harbor, NY, on 27 Feb 1815.

Bean, Peter - Recruit - 26th U.S. Infantry - Recruiting Detachment: William Bezeau - Age: 23 - Height: 5' 3" - Eyes: Brown - Hair: Curly - Complexion: Yellow - Born: Pennsylvania - Trade: Laborer - Enlistment date: 7 Nov 1814 - Enlistment place: Philadelphia - Enlistment period: 5 Yrs - Enlisted by whom: William Bezeau - Discharge date: 20 May 1815 - Comments: Listed as "Col'd" in his service record.

Bey, Levi - Private - U.S. Light Artillery - Company: John Bell - Age: 21 - Height: 5' 2 3/4" - Eyes: Black - Hair: Black - Complexion: Black - Born: CT, Hartford County, Suffield - Trade: Mariner - Enlistment date: 28 Sep 1814 - Enlistment period: War - Enlisted by whom: William Campbell - Discharge date: 31 Mar 1815 - Comments: Possible Black, not verified.

Blackford, Isaac - Recruit - 26th U.S. Infantry - Recruiting Detachment: William Bezeau - Age: 26 - Height: 5' 7" - Eyes: Black - Hair: Curly - Complexion: Black - Born: Pennsylvania - Trade: Laborer - Enlistment date: 15 Oct 1814 - Enlistment place: Philadelphia - Enlistment period: 5 Yrs - Enlisted by whom: William Bezeau - Discharge date: 23 May 1815 - Comments: Listed as "Colored" in his service record; discharged at Philadelphia, PA.

Blackston, David - Recruit - 26th U.S. Infantry - Recruiting Detachment: William Bezeau - Age: 23 - Height: 5' 5" - Eyes: Black - Hair: Curly - Complexion: Black or Chestnut - Born: Delaware or PA, Milford - Trade: Laborer - Enlistment date: 13 Oct 1814 - Enlistment place: Philadelphia - Enlistment period: 5 Yrs - Enlisted by whom: William Bezeau - Discharge date: 20 Feb 1815 - Comments: Listed as a Negro in his service record; discharged by Surgeon's Certificate for Disability at Province Island Barracks, PA.

Blake, William - Recruit - 26th U.S. Infantry - Recruiting Detachment: William Bezeau - Age: 23 - Height: 5' 7" - Eyes: Black - Hair: Black - Complexion: Black - Born: Pennsylvania - Trade: Laborer - Enlistment date: 5 Nov 1814 - Enlistment place: Philadelphia - Enlistment period: War - Enlisted by whom: William Bezeau - Comments: Listed as "Col'd" in his service record.

Bosler, John - Recruit (Sergeant) - 26th U.S. Infantry - Recruiting Detachment: William Bezeau - Age: 40 - Height: 5' 10" - Born: Pennsylvania - Enlistment date: 2 Sep 1814 - Enlistment place: Philadelphia - Enlistment period: War - Enlisted by whom: William Bezeau - Comments: Listed as "Colored" in his service record.

Bowen, John - Private - U.S. Light Artillery - Company: George Morris - Age: 27 - Height: 5' 5" - Eyes:

Black - Hair: Black - Complexion: Black - Born: MA, North Hampton or Greenbush, NY - Trade: Farmer - Enlistment date: 6 Oct 1814 - Enlistment place: MA, Boston - Enlistment period: War - Enlisted by whom: William Campbell - Discharge date: 31 Mar 1815 - Comments: Possible Black, not verified.

Bowlin, Richard - Laborer - U.S. Ordnance Department - Company: Robert Richardson - Age: 22 - Height: 6' 1" - Eyes: Yellow - Hair: Black - Complexion: Yellow - Born: PA, Cumberland County, Shippensburg - Trade: Laborer - Enlistment date: 10 Feb 1815 - Enlistment place: OH, Franklin County, Franklinton - Enlistment period: 5 Yrs - Enlisted by whom: Robert Richardson - Comments: Possible Black, not verified.

Boyington, Richard - Waiter - 11th U.S. Infantry - Company: Staff - Other regiment: 4th U.S. Infantry - Age: 16 - Height: 5' 6" - Eyes: Black - Hair: Black - Complexion: not stated - Born: Vermont - Enlistment date: 25 Jun 1812 - Enlistment period: 5 Yrs - Enlisted by whom: Malachi Corning - Discharge date: 15 May 1815 - Bounty: BLW 15954-160-12 (Illinois) - Comments: Listed as "Black" in his service records; waiter to Colonel Henry Atkinson; discharged at Philadelphia, PA, under an order from the War Department, directing all soldiers of color should be discharged.

Bracken, Hugh - Recruit - 26th U.S. Infantry - Recruiting Detachment: William Bezeau - Age: 27 - Height: 5' 9 1/2" - Eyes: Blue - Hair: Sandy - Complexion: Light - Born: Ireland - Trade: Tobacconist - Enlistment date: 2 Jul 1814 - Enlistment place: Philadelphia - Enlistment period: War - Enlisted by whom: William Bezeau - Comments: Listed as "Colored" with a question mark in his service record; deserted 24 Jul 1814.

Bradish, William - Private - 31st U.S. Infantry - Company: Andrew Arnolds - Age: 25 - Height: 5' 8 1/2" - Eyes: Black - Hair: Black - Complexion: Black - Born: NH, Hillsborough County, Litchfield - Trade: Farmer - Enlistment date: 3 Mar 1814 - Enlistment place: VT, Montpelier - Enlistment period: War - Enlisted by whom: John Hatch - Discharge date: 6 Jun 1815 - Comments: Possible Black, not verified; discharged at Plattsburgh, NY.

Bridget, Samuel - Private - 26th U.S. Infantry - Recruiting Detachment: William Bezeau - Age: 28 - Height: 5' 4 1/2" - Eyes: Black - Hair: Black - Complexion: Brown - Born: Pennsylvania - Trade: Barber - Enlistment date: 11 Nov 1814 - Enlistment place: Philadelphia - Enlistment period: 5 Yrs - Enlisted by whom: William Bezeau - Discharge date: 17 May 1815 - Comments: Listed as "Col'd" in his service record; discharged at Philadelphia, PA.

Broadwell, John - Private - 30th U.S. Infantry - Company: Gideon Spencer - Age: 18 - Height: 5' 8" - Eyes: Black - Hair: Curly - Complexion: Black - Born: New York - Trade: Farmer - Enlistment date: 22 Mar 1814 - Enlistment period: War - Enlisted by whom: William Barny - Comments: Possible Black, not verified; on duty with the 15th Infantry as a waiter.

Brown, Francis - Private - U.S. Light Artillery - Company: John Bell - Age: 24 - Height: 5' 5" - Eyes: Black - Hair: Black - Complexion: Black - Born: RI, Providence County, Providence - Enlistment date: 22 Aug 1814 - Enlistment period: War - Enlisted by whom: William Campbell - Comments: Possible Black, not verified.

Brown, George - Recruit - 26th U.S. Infantry - Recruiting Detachment: William Bezeau - Enlistment date: 6 Dec 1814 - Enlistment period: War - Comments: Listed as "Colored" in his service record.

Brown, George - Recruit - 26th U.S. Infantry - Recruiting Detachment: William Bezeau - Age: 25 - Height: 5' 6" - Eyes: Black - Hair: Curly - Complexion: Black - Born: Pennsylvania - Trade:

Laborer - Enlistment date: 4 Dec 1814 - Enlistment place: Philadelphia - Enlistment period: 5 Yrs - Enlisted by whom: William Bezeau - Discharge date: 20 May 1815 - Comments: Listed as a Negro in his service record; discharged at Philadelphia, PA.

Brown, George - Recruit - 26th U.S. Infantry - Recruiting Detachment: William Bezeau - Age: 25 - Height: 5' 7" - Eyes: Black - Hair: Curly - Complexion: Black - Born: Delaware - Trade: Laborer - Enlistment date: 2 Dec 1814 - Enlistment place: Philadelphia - Enlistment period: 5 Yrs - Enlisted by whom: William Bezeau - Discharge date: 29 May 1815 - Comments: Listed as a Negro in his service record; discharged at Philadelphia, PA.

Brown, James - Private - 1st U.S. Infantry - Company: William Whistler - Enlistment date: 17 Nov 1812 - Enlistment place: PA, Erie - Discharge date: 31 Jan 1813 - Pension: WO-285 Rejected: wife Elizabeth Dorsey, served in Captain Whistler's Company, 1st Infantry - Comments: Listed as "Colored" in the widow's pension application; died in Aug 1857 in Erie, PA; service record has not been found.

Brown, Thomas - Recruit - 26th U.S. Infantry - Recruiting Detachment: William Bezeau - Age: 24 - Height: 5' 8" - Eyes: Black - Hair: Curly - Complexion: Chestnut - Born: Maryland - Trade: Laborer - Enlistment date: 8 Oct 1814 - Enlistment place: Philadelphia - Enlistment period: War - Enlisted by whom: William Bezeau - Discharge date: 23 Mar 1815 - Bounty: BLW 25756-160-12 (Arkansas) - Comments: Listed as "Col'd" in his service record; discharged at Philadelphia, PA.

Brown, William - Private - U.S. Light Artillery - Company: John Bell - Age: 35 - Height: 5' 7" - Eyes: Black - Hair: Black - Complexion: Black - Born: Danish West Indies, St. Croix - Trade: Mariner - Enlistment date: 4 Oct 1814 - Enlistment place: MA, Boston - Enlistment period: War - Enlisted by whom: William Campbell - Discharge date: 31 Mar 1815 - Comments: Possible Black, not verified.

Buck, William - Recruit - 26th U.S. Infantry - Recruiting Detachment: William Bezeau - Age: 31 - Height: 5' 5" - Eyes: Black - Hair: Black - Complexion: Dark - Born: Pennsylvania - Trade: Laborer - Enlistment date: 21 Sep 1814 - Enlistment place: Philadelphia - Enlistment period: War - Enlisted by whom: William Bezeau - Discharge date: 23 Mar 1815 - Bounty: BLW 7181-160-12 (Illinois) - Comments: Listed as "Col'd" in his service record; discharged at Philadelphia, PA.

Burk, William - Recruit - 26th U.S. Infantry – Recruiting Detachment: William Bezeau - Enlistment date: 7 Oct 1814 - Enlistment period: War - Comments: Listed as "Col'd" in his service record.

Burkess, Jacob - Recruit - 26th U.S. Infantry - Recruiting Detachment: William Bezeau - Age: 24 - Height: 5' 7 1/2" - Eyes: Grey - Hair: Black - Complexion: Sallow - Born: Pennsylvania - Trade: Blacksmith - Enlistment date: 30 Sep 1814 - Enlistment place: Philadelphia - Enlistment period: 5 Yrs - Enlisted by whom: William Bezeau - Discharge date: 16 May 1815 - Comments: Listed as a "Colored man" in his service record; discharged at Philadelphia, PA.

Burkess, William - Recruit - 26th U.S. Infantry - Recruiting Detachment: William Bezeau - Age: 21 - Height: 5' 6" - Eyes: Black - Hair: Black - Complexion: Black - Born: Pennsylvania - Trade: Laborer - Enlistment date: 7 Oct 1814 - Enlistment place: Philadelphia - Enlistment period: War - Enlisted by whom: William Bezeau - Comments: Listed as "Col'd" in his service record; deserted on 15 Oct 1814.

Burns, Charles - Private - U.S. Light Artillery - Age: 24 - Height: 5' 3 1/2" - Eyes: Black - Hair: Black - Complexion: Black - Born: MA, Suffolk County, Boston - Trade: Rope maker - Enlistment date:

15 Jul 1814 - Enlistment place: MA, Boston - Enlistment period: War - Enlisted by whom: Samuel Washburn - Discharge date: 31 May 1815 - Comments: Listed as a "Colored man" in his service record; dishonorably discharged at Plattsburgh, NY, due to desertions.

Butler, Stephen - Private - U.S. Light Artillery - Comments: Listed as "Col'd" in his service record; court-martialed for desertion.

Cage – see Shannon Peter Caze

Calvin, Caleb - Private - 15th U.S. Infantry - Company: Joseph Barton - Age: 30 - Height: 5' 10 1/2" - Eyes: Black - Hair: Black - Complexion: Yellow - Born: NJ, Burlington County, Evesham - Trade: Laborer - Enlistment date: 6 Mar 1813 - Enlistment place: NJ, Mount Holly - Enlistment period: 5 Yrs - Enlisted by whom: Joseph Barton - Discharge date: 1 May 1815 - Bounty: BLW 26452-160-12 (Arkansas) - Comments: Possible Black, not verified: discharged at Greenbush, NY; bounty land given to Bartholomew Calvin, father & heir at law of Caleb Calvin, 10 Mar 1832.

Campbell, John - Recruit - 26th U.S. Infantry - Recruiting Detachment: William Bezeau - Age: 25 - Height: 5' 8 1/2" - Eyes: Grey - Hair: Black - Complexion: Yellow - Born: PA, Philadelphia - Trade: Tanner - Enlistment date: 13 Sep 1814 - Enlistment place: Philadelphia - Enlistment period: 5 Yrs - Enlisted by whom: William Bezeau - Discharge date: 16 May 1815 - Comments: Listed as a "Black man" in his service record; discharged at Philadelphia, PA.

Caze, Shannon Peter - Recruit - 26th U.S. Infantry - Recruiting Detachment: William Bezeau - Age: 26 - Height: 5' 8 1/2" - Eyes: Black - Hair: Black - Complexion: Black - Born: Saint Domingo (Haiti) - Trade: Farmer - Enlistment date: 11 Sep 1814 - Enlistment place: Philadelphia - Enlistment period: War - Enlisted by whom: William Bezeau - Discharge date: 23 Mar 1815 - Bounty: BLW 6440-160-12 (Illinois) - Comments: Listed as a "Colored man" in his service record; discharged at Philadelphia, PA.

Chance, Peter - Recruit - 26th U.S. Infantry - Recruiting Detachment: William Bezeau - Age: 27 - Height: 5' 5" - Eyes: Black - Hair: Black - Complexion: Chestnut - Born: Maryland - Trade: Laborer - Enlistment date: 7 Nov 1814 - Enlistment place: Philadelphia - Enlistment period: 5 Yrs - Enlisted by whom: William Bezeau - Discharge date: 18 May 1815 - Comments: Listed as a "Colored man" in his service record.

Charles, Henry - Recruit - 26th U.S. Infantry - Recruiting Detachment: William Bezeau - Age: 22 - Height: 5' - Eyes: Black - Hair: Black - Complexion: Black - Born: Pennsylvania - Trade: Farmer - Enlistment date: 11 Oct 1814 - Enlistment place: Philadelphia - Enlistment period: 5 Yrs - Enlisted by whom: William Bezeau - Discharge date: 20 May 1815 - Comments: Listed as a "Colored man" in his service record.

Chesnut, Kinsheon - Private - 10th U.S. Infantry - Company: Robert Mitchell - Other regiment: 8th U.S. Infantry - Age: 18 - Height: 5' 6" - Eyes: Black - Hair: Black - Complexion: Yellow - Born: NC, Sampson County - Trade: Laborer - Enlistment date: 9 Jun 1813 - Enlistment place: NC, Duplin County - Enlistment period: 5 Yrs - Enlisted by whom: James Hill - Discharge date: 9 Jun 1818 - Comments: Possible Black, not verified; discharged at Fort Crawford, IL.

Clinton, Alexander - Recruit - 26th U.S. Infantry - Recruiting Detachment: William Bezeau - Age: 28 - Height: 6' - Eyes: Black - Hair: Curly - Complexion: Black - Born: Virginia - Trade: Farmer - Enlistment date: 4 Jan 1815 - Enlistment place: Philadelphia - Enlistment period: War - Enlisted by whom: William Bezeau - Comments: Listed on the Descriptive Roll of Colored Men, 1 Apr 1815.

Cogler, Josiah - Recruit - 26th U.S. Infantry - Recruiting Detachment: William Bezeau - Age: 27 - Height: 5' 4" - Eyes: Black - Hair: Curly - Complexion: Black - Born: Virginia - Trade: Farmer - Enlistment date: 6 Jan 1815 - Enlistment place: Philadelphia - Enlistment period: War - Enlisted by whom: William Bezeau - Comments: Listed on the Descriptive Roll of Colored Men, 1 Apr 1815.

Comer – see Robert Coomer

Congo, Aaron - Recruit - 26th U.S. Infantry - Recruiting Detachment: William Bezeau - Age: 25 - Height: 5' 6" - Eyes: Black - Hair: Black - Complexion: Chestnut - Born: New Jersey - Trade: Laborer - Enlistment date: 16 Sep 1814 - Enlistment place: Philadelphia - Enlistment period: War - Enlisted by whom: William Bezeau - Discharge date: 20 Mar 1815 - Bounty: BLW 10494-160-12 (Illinois) - Comments: Listed as a "Colored man" in his service record; discharged at Philadelphia, PA.

Congo, James - Recruit - 26th U.S. Infantry - Recruiting Detachment: William Bezeau - Age: 26 - Height: 5' - Eyes: Black - Hair: Black - Complexion: Chestnut - Born: New Jersey - Trade: Laborer - Enlistment date: 16 Sep 1814 - Enlistment place: Philadelphia - Enlistment period: War - Enlisted by whom: William Bezeau - Discharge date: 23 Mar 1815 - Bounty: BLW 10495-160-12 (Illinois) - Comments: Listed as "Colored" in his service record; discharged at Philadelphia, PA.

Conner, Hosea - Recruit - 26th U.S. Infantry - Recruiting Detachment: William Bezeau - Age: 27 - Height: 5' 10" - Eyes: Black - Hair: Black - Complexion: Chestnut - Born: Pennsylvania - Trade: Laborer - Enlistment date: 31 Oct 1814 - Enlistment place: Philadelphia - Enlistment period: 5 Yrs - Enlisted by whom: William Bezeau - Discharge date: 18 May 1815 - Comments: Listed as a "Colored man" in his service record; discharged at Philadelphia, PA.

Conover, Richard - Recruit - 26th U.S. Infantry - Recruiting Detachment: William Bezeau - Age: 16 - Height: 5' 3 1/2" - Eyes: Hazel - Hair: Black - Complexion: Chestnut - Born: New York - Trade: Laborer - Enlistment date: 5 Oct 1814 - Enlistment place: Philadelphia - Enlistment period: 5 Yrs - Enlisted by whom: William Bezeau - Comments: Listed as "Colored" in his service record; deserted on 15 Nov 1814.

Conrad, James (Conrod) - Recruit (Sergeant) - 26th U.S. Infantry - Recruiting Detachment: William Bezeau - Age: 33 - Height: 5' 5 3/4" - Eyes: Blue - Hair: Brown - Complexion: Dark - Born: Germany - Trade: Baker - Enlistment date: 17 Aug 1814 - Enlistment place: Philadelphia - Enlistment period: War - Enlisted by whom: William Bezeau - Bounty: BLW 12419-160-12 (Illinois) - Comments: Listed as a "Colored man" in his service record; discharged at Philadelphia, PA.

Cook, George - Recruit - 26th U.S. Infantry - Recruiting Detachment: William Bezeau - Age: 24 - Height: 5' 10" - Eyes: Black - Hair: Black - Complexion: Yellow - Born: Delaware - Trade: Sailor - Enlistment date: 28 Nov 1814 - Enlistment place: Philadelphia - Enlistment period: War - Enlisted by whom: William Bezeau - Comments: Listed as "Colored" in his service record.

Coomer, Robert - Private - U.S. Light Artillery - Company: George Morris - Age: 21 - Height: 5' 5" - Eyes: Black - Hair: Black - Complexion: Black - Born: Jersey - Trade: Mariner - Enlistment date: 6 Oct 1814 - Enlistment place: MA, Boston - Enlistment period: War - Enlisted by whom: William Campbell - Comments: Listed as "Blk" in his service records; deserted on 24 Nov 1814 while on furlough.

Coone, Abraham - Recruit - 26th U.S. Infantry - Recruiting Detachment: William Bezeau - Age: 19 - Height: 5' 2" - Eyes: Black - Hair: Curly - Complexion: Black - Born: Pennsylvania - Trade: Farmer - Enlistment date: 6 Jan 1815 - Enlistment place: Philadelphia - Enlistment period: War - Enlisted by whom: William Bezeau - Comments: Listed on the Descriptive Roll of Colored Men, 1 Apr 1815.

Cooper, John - Recruit - 26th U.S. Infantry - Recruiting Detachment: William Bezeau - Age: 23 - Height: 5' 9" - Eyes: Black - Hair: Black - Complexion: Black or Chestnut - Born: Massachusetts - Trade: Laborer - Enlistment date: 9 Nov 1814 - Enlistment place: Philadelphia - Enlistment period: War - Enlisted by whom: William Bezeau - Discharge date: 23 Mar 1815 - Bounty: BLW 7185-160-12 (Illinois) - Comments: Listed as a "Colored man" in his service record; discharged at Philadelphia, PA.

Cooper, Moses - Recruit - 26th U.S. Infantry - Recruiting Detachment: William Bezeau - Age: 21 - Height: 5' 7" - Eyes: Black - Hair: Curly - Complexion: Brown - Born: Delaware - Trade: Laborer - Enlistment date: 9 Jan 1815 - Enlistment place: Philadelphia - Enlistment period: War - Enlisted by whom: William Bezeau - Comments: Listed as a "Black man" in his service record.

Cox, Joshua (Coxe) - Recruit - 26th U.S. Infantry - Recruiting Detachment: William Bezeau - Age: 21 - Height: 5' 4" - Eyes: Black - Hair: Black - Complexion: Black - Born: PA, Philadelphia - Trade: Farmer - Enlistment date: 13 Sep 1814 - Enlistment place: Philadelphia - Enlistment period: War - Enlisted by whom: William Bezeau - Discharge date: 20 Mar 1815 - Comments: Listed as a "Colored man" in his service record; discharged at Philadelphia, PA.

Crawford, H. - Private Servant - 2nd Military District - Company: General Staff - Enlistment date: 1 Jun 1814 - Comments: Black servant to Brigadier General Thomas Cushing.

Crunkilton, Charles (Crunkitton) - Recruit - 26th U.S. Infantry - Recruiting Detachment: William Bezeau - Age: 24 - Height: 5' 8 1/2" - Eyes: Hazel - Hair: Curly - Complexion: Black - Born: Delaware - Trade: Laborer - Enlistment date: 20 Dec 1814 - Enlistment place: Philadelphia - Enlistment period: War - Enlisted by whom: William Bezeau - Discharge date: 20 Mar 1815 - Bounty: BLW 535-320-14 (Illinois) - Comments: Listed as a "Colored man" in his service record; discharged at Philadelphia, PA.

Cruse, Francis L. - Private - 8th U.S. Infantry - Company: Charles Crawford - Age: 36 - Height: 5' 2 1/2" - Eyes: Black - Hair: Black - Complexion: Black - Born: Cuba, Havana - Trade: Baker - Enlistment date: 3 Dec 1812 - Enlistment place: GA, Milledgeville - Enlistment period: 5 Yrs - Enlisted by whom: Charles Crawford - Discharge date: 20 Aug 1815 - Comments: Possible Black, not verified; discharged at Fort Hawkins, GA, on disability; alias Francis La Cruse.

Culley, Charles - Private - 26th U.S. Infantry - Recruiting Detachment: William Bezeau - Age: 21 - Height: 5' 6" - Eyes: Black - Hair: Curly - Complexion: Black - Born: Delaware - Trade: Farmer - Enlistment date: 4 Feb 1815 - Enlistment place: Philadelphia - Enlistment period: War - Enlisted by whom: William Bezeau - Comments: Listed on the Descriptive Roll of Colored Men, 1 Apr 1815.

Cunnal, William - Recruit - 26th U.S. Infantry - Recruiting Detachment: William Bezeau - Age: 29 - Height: 5' 9" - Eyes: Black - Hair: Curly - Complexion: Black - Born: Pennsylvania - Trade: Farmer - Enlistment date: 2 Jan 1815 - Enlistment place: Philadelphia - Enlistment period: War - Enlisted by whom: William Bezeau - Comments: Listed on the Descriptive Roll of Colored Men, 1 Apr 1815.

Currell, Bryan - Private - 39th U.S. Infantry - Company: George Hallum - Enlistment date: 20 Dec 1813 - Discharge date: 19 Dec 1814 - Comments: Reported as a negro on the regimental muster rolls completed after the Battle of Horseshoe Bend.

Custis, George - Private - U.S. Light Artillery - Company: John Bell - Age: 35 - Height: 5' 10" - Eyes: Black - Hair: Black - Complexion: Black - Born: PA, Philadelphia - Trade: Sailor - Enlistment date: 10 Oct 1814 - Enlistment period: War - Enlisted by whom: William Campbell - Comments: Possible Black, not verified.

Cutler, Simon - Recruit - 26th U.S. Infantry - Recruiting Detachment: William Bezeau - Age: 20 - Height: 5' 6" - Eyes: Black - Hair: Curly - Complexion: Black - Born: Virginia - Enlistment date: 9 Jan 1815 - Enlistment place: Philadelphia - Enlistment period: War - Enlisted by whom: William Bezeau - Comments: Listed on the Descriptive Roll of Colored Men, 1 Apr 1815.

Cutting, Lewis - Recruit - 26th U.S. Infantry - Recruiting Detachment: William Bezeau - Age: 21 - Height: 5' 3" - Eyes: Black - Hair: Curly - Complexion: Brown - Born: Maryland - Trade: Laborer - Enlistment date: 4 Jan 1815 - Enlistment place: Philadelphia - Enlistment period: War - Enlisted by whom: William Bezeau - Comments: Listed on the Descriptive Roll of Colored Men, 1 Apr 1815.

Dana, Daniel C. - Recruit - 26th U.S. Infantry - Recruiting Detachment: William Bezeau - Age: 19 - Height: 5' 8" - Eyes: Black - Hair: Curly - Complexion: Black - Born: Pennsylvania - Trade: Shoemaker - Enlistment date: 12 Nov 1814 - Enlistment place: Philadelphia - Enlistment period: War - Enlisted by whom: William Bezeau - Comments: Listed as "Col'd" in his service record; deserted on 20 Nov 1814.

Daton, Andrew - Private - 22nd U.S. Infantry - Other regiment: 2nd U.S. Infantry - Age: 21 - Height: 5' 6" - Eyes: Dark - Hair: Dark - Complexion: Dark - Born: New Jersey - Trade: Farmer - Enlistment date: 13 Feb 1815 - Enlisted by whom: Claudius Legrand - Discharge date: 13 Jun 1815 - Comments: Listed as "Colored" in his service record; discharged at Philadelphia, PA.

Dawson, Levi - Recruit - 26th U.S. Infantry - Recruiting Detachment: William Bezeau - Age: 27 - Height: 5' 7" - Eyes: Black - Hair: Curly - Complexion: Chestnut - Born: New Jersey - Trade: Laborer - Enlistment date: 28 Dec 1814 - Enlistment place: Philadelphia - Enlistment period: War - Enlisted by whom: William Bezeau - Discharge date: 20 Mar 1815 - Bounty: BLW 148-320-14 Cancelled - Comments: Listed as "Colored" in his service record; discharged at Philadelphia, PA; bounty cancelled by an order from the War Department on 19 May 1820.

De Cass John - Private - U.S. Light Artillery - Company: George Morris - Age: 22 - Height: 5' 2 1/2" - Eyes: Black - Hair: Black - Complexion: Black - Born: Danish West Indies, St. Croix - Trade: Seaman - Enlistment date: 4 Aug 1814 - Enlistment place: MA, Boston - Enlistment period: War - Enlisted by whom: Samuel Washburn - Comments: Possible Black, not verified; deserted while on a pass, 6 Nov 1814.

Delaurier – see Baptiste, John

Deleva, Antonia - Recruit - 26th U.S. Infantry - Recruiting Detachment: William Bezeau - Age: 34 - Height: 5' 8" - Eyes: Black - Hair: Curly - Complexion: Black - Born: St. Jago (Cuba) - Trade: Seaman - Enlistment date: 10 Jan 1815 - Enlistment place: Philadelphia - Enlistment period: War - Enlisted by whom: William Bezeau - Comments: Listed on the Descriptive Roll of Colored Men, 1 Apr 1815.

Delster, James - Recruit - 26th U.S. Infantry - Recruiting Detachment: William Bezeau - Age: 24 - Height: 5' 6" - Eyes: Black - Hair: Curly - Complexion: Black - Born: Cuba - Trade: Seaman - Enlistment date: 2 Jan 1815 - Enlistment place: Philadelphia - Enlistment period: War - Enlisted by whom: William Bezeau - Bounty: BLW 831-320-14 Cancelled - Comments: Listed on the Descriptive Roll of Colored Men, 1 Apr 1815; bounty cancelled by an order from the War Department on 19 May 1820.

Dennis, John - Recruit - 26th U.S. Infantry - Recruiting Detachment: William Bezeau - Age: 28 - Height: 5' 10 1/2" - Eyes: Black - Hair: Black - Complexion: Yellow - Born: Pennsylvania - Trade: Fiddler - Enlistment date: 7 Nov 1814 - Enlistment place: Philadelphia - Enlistment period: 5 Yrs - Enlisted by whom: William Bezeau - Discharge date: 18 May 1815 - Comments: Listed as a Negro in his service record; discharged at Philadelphia, PA.

Derbey, Abraham - Laborer - Corps of Artificers - Company: Alexander Paris - Comments: Listed as "Black" in *A List of Pensioners of the War of 1812*.

Derwood, Joshua - Recruit - 26th U.S. Infantry - Recruiting Detachment: William Bezeau - Age: 27 - Height: 5' 10 1/2" - Eyes: Black - Hair: Curly - Complexion: Black - Born: New Jersey - Trade: Farmer - Enlistment date: 1 Feb 1815 - Enlistment place: Philadelphia - Enlistment period: War - Enlisted by whom: William Bezeau - Comments: Listed on the Descriptive Roll of Colored Men, 1 Apr 1815.

Dexter, Jacob - Private - 25th U.S. Infantry - Company: Daniel Ketchum - Height: 5' 9" - Eyes: Black - Hair: Black - Complexion: Black - Trade: Farmer - Enlistment date: 18 Mar 1814 - Enlistment place: CT, Windsor - Enlistment period: War - Enlisted by whom: Ephraim Shaylor - Discharge date: 28 Mar 1815 - Bounty: BLW 25792-160-12 (Arkansas) - Comments: Listed as a "Colored man" in his service record; wounded in the Battle of Chippewa, UC, 5 Jul 1814.

Dickinson, Harry - Private - 30th U.S. Infantry - Company: William Miller - Other regiment: Corps of Artillery - Age: 21 - Height: 5' - Eyes: Black - Hair: Curly or Black - Complexion: Black - Born: PA, Philadelphia County, Philadelphia - Trade: Cordwainer - Enlistment date: 8 Apr 1814 - Enlistment place: VT, Burlington - Enlistment period: War - Enlisted by whom: William Barney - Comments: Possible Black, not verified.

Domingo, Emanuel - Private - U.S. Light Artillery - Company: George Morris - Age: 26 - Height: 5' 9" - Eyes: Black - Hair: Black - Complexion: Black - Born: Danish West Indies, St. Croix - Trade: Seaman - Enlistment date: 8 Aug 1814 - Enlistment place: MA, Boston - Enlistment period: War - Enlisted by whom: Samuel Washburn - Discharge date: 31 Mar 1815 - Comments: Possible Black, not verified; also listed as Manuel Domingo.

Doras, John - Recruit - U.S. Light Artillery - Age: 19 - Height: 5' 3 3/4" - Born: Saint Domingo (Haiti) - Enlistment date: 22 Jul 1814 - Enlistment place: MA, Boston - Enlisted by whom: Samuel Washburn - Comments: Possible Black, not verified.

Dorsey, Charles - Recruit - 26th U.S. Infantry - Recruiting Detachment: William Bezeau - Age: 38 - Height: 6' - Eyes: Black - Hair: Curly - Complexion: Chestnut - Born: Pennsylvania - Trade: Farmer - Enlistment date: 10 Feb 1815 - Enlistment place: Philadelphia - Enlistment period: War - Enlisted by whom: William Bezeau - Bounty: BLW 810-320-14 Cancelled; BLW 1036-320-14 - Comments: Listed on the Descriptive Roll of Colored Men, 1 Apr 1815; land bounty to Mathew Dorsey, brother & only heir at law of Charles Dorsey; he died before 6 Jul 1819; first bounty cancelled by an order from the War Department on 19 May 1820.

Draper, Samuel - Recruit - 26th U.S. Infantry - Recruiting Detachment: William Bezeau - Age: 19 - Height: 5' 6" - Eyes: Black - Hair: Black - Complexion: Black - Born: Maryland - Trade: Laborer - Enlistment date: 11 Oct 1814 - Enlistment place: Philadelphia - Enlistment period: 5 Yrs - Enlisted by whom: William Bezeau - Comments: Listed as "Col'd" in his service record; discharged for being a minor.

Drummond, Levin (also Levi) - Recruit - 26th U.S. Infantry - Recruiting Detachment: William Bezeau - Age: 25 - Height: 5' 6" - Eyes: Black - Hair: Curly - Complexion: Brown - Born: Virginia - Trade: Farmer - Enlistment date: 17 Nov 1814 - Enlistment place: Philadelphia - Enlistment period: War - Enlisted by whom: William Bezeau - Discharge date: 20 Mar 1815 - Bounty: BLW 7187-160-12 (Illinois) - Comments: Listed as "Col'd" in his service record; discharged at Philadelphia, PA.

Duffin, Edward - Recruit - 26th U.S. Infantry - Recruiting Detachment: William Bezeau - Age: 26 - Height: 5' 8" - Eyes: Black - Hair: Curly - Complexion: Black - Born: Pennsylvania - Trade: Laborer - Enlistment date: 3 Jan 1815 - Enlistment place: Philadelphia - Enlistment period: War - Enlisted by whom: William Bezeau - Comments: Listed on the Descriptive Roll of Colored Men, 1 Apr 1815.

Duke, Simon - Recruit - 26th U.S. Infantry - Recruiting Detachment: William Bezeau - Age: 21 - Height: 5' 4" - Eyes: Black - Hair: Curly - Complexion: Black - Born: Pennsylvania - Trade: Laborer - Enlistment date: 3 Feb 1815 - Enlistment place: Philadelphia - Enlistment period: War - Enlisted by whom: William Bezeau - Comments: Listed on the Descriptive Roll of Colored Men, 1 Apr 1815.

Dunaway, William - Recruit - 26th U.S. Infantry - Recruiting Detachment: William Bezeau - Age: 21 - Height: 5' - Eyes: Black - Hair: Curly - Complexion: Black - Born: Massachusetts - Trade: Laborer - Enlistment date: 11 Oct 1814 - Enlistment place: Philadelphia - Enlistment period: 5 Yrs - Enlisted by whom: William Bezeau - Discharge date: 20 May 1815 - Comments: Listed as "Col'd" in his service record; discharged at Philadelphia, PA.

Durang, J. M. F. A. - Recruit - U.S. Light Artillery - Age: 29 - Born: Isle of France (Mauritius) - Enlistment date: 8 Nov 1814 - Enlistment place: RI, Providence - Enlisted by whom: Benjamin Robb - Comments: Possible Black, not verified.

Durker, Alexander - Recruit - 26th U.S. Infantry - Recruiting Detachment: William Bezeau - Age: 27 - Height: 5' 4" - Eyes: Black - Hair: Curly - Complexion: Brown - Born: New York - Trade: Farmer - Enlistment date: 1 Jan 1815 - Enlistment place: Philadelphia - Enlistment period: War - Enlisted by whom: William Bezeau - Comments: Listed on the Descriptive Roll of Colored Men, 1 Apr 1815; deserted on 9 Jan 1815.

Durkham, Vincent - Recruit - 26th U.S. Infantry - Recruiting Detachment: William Bezeau - Age: 31 - Height: 5' 9" - Eyes: Hazel - Hair: Black - Complexion: Yellow - Born: Delaware - Trade: Laborer - Enlistment date: 13 Nov 1814 - Enlistment place: Philadelphia - Enlistment period: 5 Yrs - Enlisted by whom: William Bezeau - Comments: Listed as "Colored" in his service record.

Dury, Joseph - Recruit - 26th U.S. Infantry - Recruiting Detachment: William Bezeau - Age: 27 - Height: 5' 11" - Eyes: Black - Hair: Curly - Complexion: Black - Born: New York - Trade: Laborer - Enlistment date: 1 Feb 1815 - Enlistment place: Philadelphia - Enlistment period: War - Enlisted by whom: William Bezeau - Comments: Possible Black, not verified; deserted on 6 Feb 1815.

Eames, John - Private - U.S. Light Artillery - Company: John Bell - Age: 24 - Height: 5' 2" - Eyes: Black

- Hair: Black - Complexion: Black - Born: NJ, Newtown or Middleton - Trade: Farmer - Enlistment date: 4 Nov 1814 - Enlistment place: MA, Boston - Enlistment period: War - Enlisted by whom: William Campbell - Discharge date: 31 Mar 1815 - Comments: Possible Black, not verified.

Ebant – see George Eliant

Ebrian, Charles - Recruit - 26th U.S. Infantry - Recruiting Detachment: William Bezeau - Age: 22 - Height: 5' 11" - Eyes: Black - Hair: Curly - Complexion: Black - Born: Pennsylvania - Trade: Laborer - Enlistment date: 20 Jan 1815 - Enlistment place: Philadelphia - Enlistment period: War - Enlisted by whom: William Bezeau - Comments: Listed on the Descriptive Roll of Colored Men, 1 Apr 1815.

Edderley, William - Recruit - 26th U.S. Infantry - Recruiting Detachment: William Bezeau - Age: 30 - Height: 5' 8" - Eyes: Black - Hair: Curly - Complexion: Black - Born: Pennsylvania - Trade: Laborer - Enlistment date: 3 Jan 1815 - Enlistment place: Philadelphia - Enlistment period: War - Enlisted by whom: William Bezeau - Comments: Listed on the Descriptive Roll of Colored Men, 1 Apr 1815.

Edmunds, Joshua - Recruit - 26th U.S. Infantry - Recruiting Detachment: William Bezeau - Age: 28 - Height: 5' 5" - Eyes: Black - Hair: Curly - Complexion: Black - Born: Pennsylvania - Trade: Farmer - Enlistment date: 16 Jan 1815 - Enlistment place: Philadelphia - Enlistment period: War - Enlisted by whom: William Bezeau - Comments: Listed on the Descriptive Roll of Colored Men, 1 Apr 1815.

Edwards, Simon - Recruit - 26th U.S. Infantry - Recruiting Detachment: William Bezeau - Age: 37 - Height: 5' 9" - Eyes: Black - Hair: Curly - Complexion: Black - Born: Pennsylvania - Trade: Farmer - Enlistment date: 28 Jan 1815 - Enlistment place: Philadelphia - Enlistment period: War - Enlisted by whom: William Bezeau - Comments: Listed on the Descriptive Roll of Colored Men, 1 Apr 1815.

Edwood, James - Recruit - 26th U.S. Infantry - Recruiting Detachment: William Bezeau - Age: 24 - Height: 5' 6" - Eyes: Black - Hair: Curly - Complexion: Black - Born: Kentucky - Trade: Farmer - Enlistment date: 4 Feb 1815 - Enlistment place: Philadelphia - Enlistment period: War - Enlisted by whom: William Bezeau - Comments: Listed on the Descriptive Roll of Colored Men, 1 Apr 1815.

Eliant, George - Recruit - 26th U.S. Infantry - Recruiting Detachment: William Bezeau - Age: 31 - Height: 5' 9" - Eyes: Black - Hair: Curly - Complexion: Brown - Born: Pennsylvania - Trade: Farmer - Enlistment date: 12 Feb 1815 - Enlistment place: Philadelphia - Enlistment period: War - Enlisted by whom: William Bezeau - Comments: Listed on the Descriptive Roll of Colored Men, 1 Apr 1815.

Ellitte, Moses - Recruit - 26th U.S. Infantry - Recruiting Detachment: William Bezeau - Age: 21 - Height: 5' 3" - Eyes: Black - Hair: Curly - Complexion: Chestnut - Born: Delaware - Trade: Farmer - Enlistment date: 5 Jan 1815 - Enlistment place: Philadelphia - Enlistment period: War - Enlisted by whom: William Bezeau - Comments: Listed on the Descriptive Roll of Colored Men, 1 Apr 1815.

Ellsberry, Charles (Ellsbury) - Private - U.S. Light Artillery - Company: George Morris - Age: 35 - Height: 5' 8 1/2" - Eyes: Black - Hair: Black - Complexion: Black - Born: DE, Kent County, Dover - Trade: Mariner - Enlistment date: 5 Oct 1814 - Enlistment place: MA, Boston - Enlistment period: 5 Yrs - Enlisted by whom: William Campbell - Discharge date: 31 Mar 1815 - Comments: Listed as "Black" in his service record and "discharged on account of being a negro and unfit to accompany American soldiers."

Embar, Peter - Recruit - 26th U.S. Infantry - Recruiting Detachment: William Bezeau - Age: 30 - Height: 6' - Eyes: Black - Hair: Curly - Complexion: Black - Born: Pennsylvania - Trade: Farmer - Enlistment date: 10 Jan 1815 - Enlistment place: Philadelphia - Enlistment period: War - Enlisted by whom: William Bezeau - Comments: Listed on the Descriptive Roll of Colored Men, 1 Apr 1815.

Eubern, Smith - Recruit - 26th U.S. Infantry - Recruiting Detachment: William Bezeau - Age: 27 - Height: 5' 6" - Eyes: Black - Hair: Curly - Complexion: Brown - Born: Pennsylvania - Trade: Laborer - Enlistment date: 10 Jan 1815 - Enlistment place: Philadelphia - Enlistment period: War - Enlisted by whom: William Bezeau - Comments: Listed on the Descriptive Roll of Colored Men, 1 Apr 1815; deserted on 14 Jan 1815.

Fambert, Lewis - Recruit - 26th U.S. Infantry - Recruiting Detachment: William Bezeau - Age: 22 - Height: 5' 9" - Eyes: Black - Hair: Curly - Complexion: Black - Born: Pennsylvania - Trade: Farmer - Enlistment date: 9 Feb 1815 - Enlistment place: Philadelphia - Enlistment period: War - Enlisted by whom: William Bezeau - Comments: Listed on the Descriptive Roll of Colored Men, 1 Apr 1815; deserted on 15 Feb 1815.

Farmer, Cuff - Recruit - 26th U.S. Infantry - Recruiting Detachment: William Bezeau - Age: 36 - Height: 5' 7" - Eyes: Black - Hair: Curly - Complexion: Black - Born: New Jersey - Trade: Farmer - Enlistment date: 28 Nov 1814 - Enlistment place: Philadelphia - Enlistment period: 5 Yrs - Enlisted by whom: William Bezeau - Discharge date: 30 Mar 1815 - Comments: Listed as "Col'd" in his service record.

Fedlow, Edward - Recruit - 26th U.S. Infantry - Recruiting Detachment: William Bezeau - Age: 21 - Height: 5' 7" - Eyes: Black - Hair: Curly - Complexion: Black - Born: New Jersey - Trade: Farmer - Enlistment date: 5 Jan 1815 - Enlistment place: Philadelphia - Enlistment period: War - Enlisted by whom: William Bezeau - Comments: Possible Black, not verified.

Fobden, Ezekiel - Recruit - 26th U.S. Infantry - Recruiting Detachment: William Bezeau - Age: 34 - Height: 5' 11" - Eyes: Black - Hair: Curly - Complexion: Black - Born: Pennsylvania - Trade: Farmer - Enlistment date: 8 Feb 1815 - Enlistment place: Philadelphia - Enlistment period: War - Enlisted by whom: William Bezeau - Bounty: BLW 809-320-14 Cancelled - Comments: Listed on the Descriptive Roll of Colored Men, 1 Apr 1815; land bounty to Jacob Boys, uncle & only heir of Ezekiel Fobden, deceased; died before 1 Jul 1819; bounty cancelled by an order from the War Department on 19 May 1820.

Fobler, Benjamin - Recruit - 26th U.S. Infantry - Recruiting Detachment: William Bezeau - Age: 30 - Height: 5' 4" - Eyes: Black - Hair: Curly - Complexion: Yellow - Born: Delaware - Trade: Miller - Enlistment date: 15 Nov 1814 - Enlistment place: Philadelphia - Enlistment period: 5 Yrs - Enlisted by whom: William Bezeau - Discharge date: 30 May 1815 - Bounty: BLW 3790-160-12 Cancelled - Comments: Listed on the Descriptive Roll of Colored Men, 1 Apr 1815; discharged at Philadelphia, PA; bounty cancelled by an order from the War Department on 19 May 1820.

Fobler, John - Recruit - 26th U.S. Infantry - Recruiting Detachment: William Bezeau - Age: 30 - Height: 5' 4" - Born: Delaware - Enlistment date: 15 Nov 1814 - Enlistment place: Philadelphia - Enlisted by whom: William Bezeau - Comments: Listed as "Col'd" in his service record.

Folbert, Titus - Recruit - 26th U.S. Infantry - Recruiting Detachment: William Bezeau - Age: 28 - Height: 5' 6" - Eyes: Black - Hair: Curly - Complexion: Yellow - Born: Ohio - Trade: Farmer - Enlistment date: 25 Jan 1815 - Enlistment place: Philadelphia - Enlistment period: War - Enlisted by whom: William Bezeau - Comments: Listed on the Descriptive Roll of Colored Men, 1 Apr 1815.

Ford, James - Recruit - 26th U.S. Infantry - Recruiting Detachment: William Bezeau - Age: 17 - Height: 5' 5" - Eyes: Black - Hair: Black - Complexion: Yellow - Born: Pennsylvania - Trade: Laborer - Enlistment date: 11 Oct 1814 - Enlistment place: Philadelphia - Enlistment period: War - Enlisted by whom: William Bezeau - Comments: Listed as "Col'd" in his service record; discharged for being a minor.

Forrest, Charles - Recruit - 26th U.S. Infantry - Recruiting Detachment: William Bezeau - Age: 34 - Height: 5' 7" - Eyes: Black - Hair: Curly - Complexion: Black - Born: Pennsylvania - Trade: Laborer - Enlistment date: 11 Feb 1815 - Enlistment place: Philadelphia - Enlistment period: War - Enlisted by whom: William Bezeau - Comments: Listed on the Descriptive Roll of Colored Men, 1 Apr 1815.

Fortland, Isaac - Recruit - 26th U.S. Infantry - Recruiting Detachment: William Bezeau - Age: 29 - Height: 5' 8" - Eyes: Black - Hair: Curly - Complexion: Black - Born: Pennsylvania - Trade: Laborer - Enlistment date: 26 Jan 1815 - Enlistment place: Philadelphia - Enlistment period: War - Enlisted by whom: William Bezeau - Comments: Listed on the Descriptive Roll of Colored Men, 1 Apr 1815.

Francis, Joseph - Private - U.S. Light Artillery - Company: George Morris - Age: 23 - Height: 5' 6 1/2" - Eyes: Black - Hair: Black - Complexion: Black - Born: San Salvador, Bahamas - Trade: Barber - Enlistment date: 18 Jul 1814 - Enlistment place: MA, Boston - Enlistment period: War - Enlisted by whom: Samuel Washburn - Discharge date: 31 May 1815 - Comments: Possible Black, not verified; discharged at Plattsburgh, NY.

Francy – see Jacob Freeny

Franklin, Jacob - Recruit - 26th U.S. Infantry - Recruiting Detachment: William Bezeau - Age: 35 - Height: 5' 10" - Eyes: Black - Hair: Curly - Complexion: Brown - Born: Pennsylvania - Trade: Farmer - Enlistment date: 27 Jan 1815 - Enlistment place: Philadelphia - Enlistment period: War - Enlisted by whom: William Bezeau - Comments: Listed on the Descriptive Roll of Colored Men, 1 Apr 1815.

Frederick, Lewis - Private - U.S. Light Artillery - Company: John McIntosh - Age: 25 - Height: 5' 7 1/2" - Eyes: Black - Hair: Black - Complexion: Black - Born: MD, Baltimore - Trade: Mariner - Enlistment date: 14 Jul 1814 - Enlistment place: MA, Boston - Enlistment period: War - Enlisted by whom: Samuel Washburn - Discharge date: 31 May 1815 - Comments: Possible Black, not verified; discharged at Plattsburgh, NY.

Freeman, George - Recruit - 26th U.S. Infantry - Recruiting Detachment: William Bezeau - Age: 30 - Height: 5' 8" - Eyes: Black - Hair: Curly - Complexion: Brown - Born: Pennsylvania - Trade: Laborer - Enlistment date: 21 Nov 1814 - Enlistment place: Philadelphia - Enlistment period: 5 Yrs - Enlisted by whom: William Bezeau - Discharge date: 1 May 1815 - Comments: Listed as "Col'd" in his service record; discharged at Philadelphia, PA.

Freeman, Joseph - Recruit - 26th U.S. Infantry - Recruiting Detachment: William Bezeau - Age: 26 -

Height: 5' - Eyes: Black - Hair: Black - Complexion: Yellow - Born: New Jersey - Trade: Shoemaker - Enlistment date: 14 Sep 1814 - Enlistment place: Philadelphia - Enlistment period: 5 Yrs - Enlisted by whom: William Bezeau - Discharge date: 17 May 1815 - Comments: Listed as "Col'd" in his service record; discharged at Philadelphia, PA.

Freeny, Jacob - Recruit - 26th U.S. Infantry - Recruiting Detachment: William Bezeau - Age: 30 - Height: 5' 8" - Eyes: Black - Hair: Curly - Complexion: Black - Born: Delaware - Trade: Laborer - Enlistment date: 20 Dec 1814 - Enlistment place: Philadelphia - Enlistment period: War - Enlisted by whom: William Bezeau - Comments: Listed as "Col'd" in his service record.

Fudy, G. B. - Private Servant - 2nd Military District - Company: General Staff - Enlistment date: 1 Jun 1814 - Discharge date: 30 Jun 1814 - Comments: Listed as "Black" in his service record; servant on the general staff.

Fuller, Jacob - Recruit - 26th U.S. Infantry - Recruiting Detachment: William Bezeau - Age: 26 - Height: 5' 5" - Eyes: Black - Hair: Curly - Complexion: Black - Born: Maryland - Trade: Farmer - Enlistment date: 18 Jan 1815 - Enlistment place: Philadelphia - Enlistment period: War - Enlisted by whom: William Bezeau - Comments: Listed on the Descriptive Roll of Colored Men, 1 Apr 1815; deserted on 22 Jan 1815.

Fuller, Quamenaugh - Private - U.S. Light Artillery - Company: John Bell - Bounty: BLW 12183-160-12 (Illinois) - Comments: Service record not found.

Fulting, Jeremiah (Futting) - Recruit - 26th U.S. Infantry - Recruiting Detachment: William Bezeau - Age: 34 - Height: 5' 11" - Eyes: Black - Hair: Curly - Complexion: Black - Born: Maryland - Trade: Laborer - Enlistment date: 24 Jan 1815 - Enlistment place: Philadelphia - Enlistment period: War - Enlisted by whom: William Bezeau - Bounty: BLW 832-320-14 Cancelled; BLW 1039-320-14 (Arkansas) - Comments: Listed on the Descriptive Roll of Colored Men, 1 Apr 1815; land bounty to Jeremiah Hubbard, cousin & only heir at law of Jeremiah Fulting, deceased; he died before 21 Oct 1819; first bounty was cancelled by an order from the War Department, 19 May 1820.

Gansey, William - Recruit - 26th U.S. Infantry - Recruiting Detachment: William Bezeau - Age: 30 - Height: 5' 8" - Eyes: Black - Hair: Curly - Complexion: Chestnut - Born: Africa - Trade: Laborer - Enlistment date: 16 Dec 1814 - Enlistment place: Philadelphia - Enlistment period: War - Enlisted by whom: William Bezeau - Discharge date: 20 Mar 1815 - Bounty: BLW 7190-160-12 (Illinois) - Comments: Listed as "Col'd" in his service record; discharged at Philadelphia, PA.

Gardiner, Edward - Recruit - 26th U.S. Infantry - Recruiting Detachment: William Bezeau - Age: 29 - Height: 5' 6" - Eyes: Black - Hair: Curly - Complexion: Black - Born: New Jersey - Trade: Laborer - Enlistment date: 21 Sep 1814 - Enlistment place: Philadelphia - Enlistment period: War - Enlisted by whom: William Bezeau - Discharge date: 23 May 1815 - Bounty: BLW 3789-160-12 - Comments: Listed as "Col'd" in his service record; discharged at Philadelphia, PA.

Gardner, William - Private - U.S. Light Artillery - Company: George Morris - Age: 19 - Height: 5' 5 1/2" - Eyes: Black - Hair: Black - Complexion: Black - Born: MA, Suffolk County, Boston - Trade: Laborer - Enlistment date: 29 Jul 1814 - Enlistment place: MA, Boston - Enlistment period: War - Enlisted by whom: Samuel Washburn - Discharge date: 31 Mar 1815 - Comments: Listed as "Black" in his service record.

Gardner, William (Gardiner) - Private - U.S. Light Artillery - Company: George Morris - Age: 25 -

Height: 5' 11 1/2" - Eyes: Black - Hair: Dark - Complexion: Dark - Born: NH, Rockingham County, Londonderry - Trade: Laborer - Enlistment date: 15 Aug 1814 - Enlistment place: MA, Norfolk County, Dedham - Enlistment period: War - Enlisted by whom: William Campbell - Comments: Listed as "Black" in his service record.

Garland, William - Recruit - 26th U.S. Infantry - Recruiting Detachment: William Bezeau - Age: 34 - Height: 5' 6" - Eyes: Black - Hair: Curly - Complexion: Chestnut - Born: Pennsylvania - Trade: Laborer - Enlistment date: 28 Jan 1815 - Enlistment place: Philadelphia - Enlistment period: War - Enlisted by whom: William Bezeau - Comments: Listed on the Descriptive Roll of Colored Men, 1 Apr 1815.

Garrish, Jeremiah - Private - U.S. Light Artillery - Company: John Bell - Age: 31 - Height: 5' 3 3/4" - Eyes: Black - Hair: Black or Dark - Complexion: Black or Dark - Born: NH, Hampton or Portsmouth - Trade: Mariner - Enlistment date: 30 Sep 1814 - Enlistment place: MA, Boston - Enlistment period: 5 Yrs - Enlisted by whom: William Campbell - Discharge date: 31 Mar 1815 - Comments: "Discharged on account of being a negro and unfit to accompany American soldiers."

Gates, Philip (also Phil) - Recruit - 26th U.S. Infantry - Recruiting Detachment: William Bezeau - Age: 37 - Height: 5' 10" - Eyes: Black - Hair: Black - Complexion: Yellow - Born: Pennsylvania - Trade: Weaver - Enlistment date: 23 Oct 1814 - Enlistment place: Philadelphia - Enlistment period: 5 Yrs - Enlisted by whom: William Bezeau - Discharge date: 1 May 1815 - Comments: Listed as "Col'd" in his service record; discharged at Philadelphia, PA.

George, William - Recruit - 26th U.S. Infantry - Recruiting Detachment: William Bezeau - Age: 18 - Height: 5' 5" - Eyes: Black - Hair: Curly - Complexion: Black - Born: Africa - Trade: Farmer - Enlistment date: 6 Feb 1815 - Enlistment place: Philadelphia - Enlistment period: War - Enlisted by whom: William Bezeau - Comments: Listed on the Descriptive Roll of Colored Men, 1 Apr 1815.

George, William - Recruit - 26th U.S. Infantry - Recruiting Detachment: William Bezeau - Age: 22 - Height: 5' 8" - Eyes: Black - Hair: Black - Complexion: Black - Born: Virginia - Trade: Blacksmith - Enlistment date: 7 Oct 1814 - Enlistment place: Philadelphia - Enlistment period: War - Enlisted by whom: William Bezeau - Discharge date: 20 Mar 1815 - Bounty: BLW 7191-160-12 (Illinois) - Comments: Listed as "Col'd" in his service record; discharged at Philadelphia, PA.

Gifter, Alexander - Recruit - 26th U.S. Infantry - Recruiting Detachment: William Bezeau - Age: 22 - Height: 5' 7" - Eyes: Black - Hair: Curly - Complexion: Yellow - Born: Georgia - Trade: Farmer - Enlistment date: 7 Feb 1815 - Enlistment place: Philadelphia - Enlistment period: War - Enlisted by whom: William Bezeau - Comments: Listed on the Descriptive Roll of Colored Men, 1 Apr 1815.

Gilbert, Nathan - Private - 31st U.S. Infantry - Company: Ethan Burnap - Age: 18 - Height: 5' 11 1/2" - Eyes: Black - Hair: Black - Complexion: Black - Born: MA, Suffolk County, Boston - Trade: Laborer - Enlistment date: 3 Apr 1814 - Enlistment place: VT, Thetford - Enlistment period: War - Enlisted by whom: Amos Brown - Discharge date: 4 Jun 1815 - Comments: Possible Black, not verified; discharged at Plattsburgh, NY.

Goforth, Daniel - Recruit - 26th U.S. Infantry - Recruiting Detachment: William Bezeau - Age: 28 - Height: 5' 8" - Eyes: Black - Hair: Curly - Complexion: Chestnut - Born: Pennsylvania - Trade: Laborer - Enlistment date: 1 Jan 1815 - Enlistment place: Philadelphia - Enlistment period: 5 Yrs - Enlisted by whom: William Bezeau - Discharge date: 1 May 1815 - Comments: Listed as a Negro in his service record; listed on the Descriptive Rolls of Colored Men, dated 25 Mar and 1 Apr 1815; discharged at Philadelphia, PA.

Gomaus, James (Gomans) - Private - 11th U.S. Infantry - Company: Samuel Holley - Age: 27 - Height: 5' 6" - Eyes: Black - Hair: Black - Complexion: Black - Trade: Baker - Enlistment date: 18 Jun 1812 - Enlistment period: 5 Yrs - Died: 19 Feb 1813 - Comments: Possible Black, not verified.

Gossand, Abraham - Private - 26th U.S. Infantry - Company: William Bezeau - Age: 25 - Height: 5' 6" - Eyes: Black - Hair: Curly - Complexion: Black - Born: New York - Trade: Seaman - Enlistment date: 22 Jan 1815 - Enlistment place: Philadelphia - Enlistment period: War - Enlisted by whom: William Bezeau - Comments: Listed on the Descriptive Roll of Colored Men, 1 Apr 1815.

Grashier, Francis - Private - 17th U.S. Infantry - Company: William Adair - Other regiment: 3rd U.S. Infantry - Age: 25 - Height: 5' 8" - Eyes: Black or Dark - Hair: Black or Dark - Complexion: Very Dark - Born: South America - Trade: Laborer - Enlistment date: 27 Apr 1814 - Enlistment place: KY, Louisville - Enlistment period: 5 Yrs - Enlisted by whom: Charles Querey - Discharge date: 1 Feb 1816 - Comments: Possible Black, not verified; discharged on Surgeon's Certificate of Disability for fractured shoulder.

Gray, John - Recruit - 26th U.S. Infantry - Recruiting Detachment: William Bezeau - Age: 27 - Height: 5' 7" - Eyes: Black - Hair: Black - Complexion: Brown - Born: Delaware - Trade: Laborer - Enlistment date: 6 Nov 1814 - Enlistment place: Philadelphia - Enlistment period: War - Enlisted by whom: William Bezeau - Discharge date: 30 Mar 1815 - Comments: Listed as "Col'd" in his service record; discharged because of a disability.

Grinton, Martin - Private - 10th U.S. Infantry - Company: Robert Mitchell - Age: 20 - Height: 5' 9" - Eyes: Black - Hair: Black - Complexion: Black - Born: NC, Wilkes County - Trade: Farmer - Enlistment date: 23 Apr 1813 - Enlistment place: NC, Wilkes County - Enlistment period: 5 Yrs - Enlisted by whom: Willie Gordon - Discharge date: 27 Aug 1815 - Pension: SO-27621 Rejected; served in Captains Robert Mitchell's, E. J. Leigh's and H. H. Villard's Companies, 10th Infantry - Comments: Listed as "Colored" in his service record; discharged at Elkton, MD.

Grinton, Philip - Private - 10th U.S. Infantry - Company: William Bailey - Other regiment: 8th U.S. Infantry - Age: 20 - Height: 5' 8 1/2" - Eyes: Grey - Hair: Dark - Complexion: Dark - Born: NC, Wilkes County - Trade: Farmer - Enlistment date: 12 Aug 1813 - Enlistment place: NC, Buncombe County - Enlistment period: 5 Yrs - Enlisted by whom: Willie Gordon - Died: 26 Sep 1815 - Bounty: BLW 25722-160-12 (Arkansas) - Comments: Died at Portage de Sioux, MO, of ague & fever; land bounty to Martin Grinton, brother & other heirs at law of Philip Grinton.

Grinton, Robert - Private - 10th U.S. Infantry - Company: Robert Mitchell - Age: 24 - Height: 5' 3" - Eyes: Black or dark - Hair: Black or Dark - Complexion: Dark or Yellow - Born: NC, Wilkes County, Wilksborough - Trade: Farmer - Enlistment date: 23 Apr 1813 - Enlistment place: NC, Wilkes County - Enlistment period: 5 Yrs - Enlisted by whom: Willie Gordon - Discharge date: 26 Aug 1815 - Comments: Listed as "Colored" in his service record; discharged, unfit for duty.

Hall, Frederick – see William Williams

Halliard, Cyrus - Recruit - 26th U.S. Infantry - Recruiting Detachment: William Bezeau - Age: 30 - Height: 5' 8" - Eyes: Black - Hair: Curly - Complexion: Black - Born: Pennsylvania - Trade: Farmer - Enlistment date: 23 Jan 1815 - Enlistment place: Philadelphia - Enlistment period: War - Enlisted by whom: William Bezeau - Comments: Listed on the Descriptive Roll of Colored Men, 1 Apr 1815.

Hames, John - Private - 30th US Infantry - Company: Gideon Spencer - Enlistment date: 8 Apr 1814 - Enlistment period: War - Discharge date: 3 Jun 1814 - Comments: Colored soldier; discharged for sore legs.

Haner, Peter - Recruit - 26th U.S. Infantry - Recruiting Detachment: William Bezeau - Age: 20 - Height: 5' 3" - Eyes: Black - Hair: Curly - Complexion: Black - Born: New Jersey - Trade: Farmer - Enlistment date: 26 Jan 1815 - Enlistment place: Philadelphia - Enlistment period: War - Enlisted by whom: William Bezeau - Comments: Listed on the Descriptive Roll of Colored Men, 1 Apr 1815.

Harn, James - Recruit - 26th U.S. Infantry - Recruiting Detachment: William Bezeau - Age: 27 - Height: 6' - Eyes: Black - Hair: Black - Complexion: Brown - Born: Virginia - Trade: Laborer - Enlistment date: 22 Oct 1814 - Enlistment place: Philadelphia - Enlistment period: War - Enlisted by whom: William Bezeau - Discharge date: 23 Mar 1815 - Bounty: BLW 7193-160-12 (Illinois) - Comments: Listed as "Col'd" in his service record; discharged at Philadelphia, PA.

Harriot, Titus - Recruit - 26th U.S. Infantry - Recruiting Detachment: William Bezeau - Age: 25 - Height: 5' 11" - Eyes: Black - Hair: Curly - Complexion: Black - Born: Pennsylvania - Trade: Farmer - Enlistment date: 8 Feb 1815 - Enlistment place: Philadelphia - Enlistment period: War - Enlisted by whom: William Bezeau - Comments: Listed on the Descriptive Roll of Colored Men, 1 Apr 1815.

Harriot, William - Private - 26th U.S. Infantry - Recruiting Detachment: William Bezeau - Age: 24 - Height: 5' 9 1/2" - Eyes: Black - Hair: Curly - Complexion: Brown - Born: Pennsylvania - Trade: Farmer - Enlistment date: 24 Jan 1815 - Enlistment place: Philadelphia - Enlistment period: War - Enlisted by whom: William Bezeau - Comments: Listed on the Descriptive Roll of Colored Men, 1 Apr 1815.

Harris, Samuel - Recruit - 26th U.S. Infantry - Recruiting Detachment: William Bezeau - Age: 24 - Height: 5' 10 1/2" - Eyes: Black - Hair: Black - Complexion: Black - Born: New York - Trade: Farmer - Enlistment date: 10 Oct 1814 - Enlistment place: Philadelphia - Enlistment period: 5 Yrs - Enlisted by whom: William Bezeau - Discharge date: 7 May 1815 - Comments: Listed as a Negro and as "Colored" in his service record; discharged at Philadelphia, PA, Surgeon's Certificate for Disability, unfit for service.

Harris, William - Private - 32nd U.S. Infantry - Age: 26 - Height: 5' 5 1/2" - Eyes: Negro - Hair: Negro - Complexion: Negro - Born: NY, Long Island - Trade: Laborer - Enlistment date: 3 Feb 1815 - Enlisted by whom: Claudius La Grand - Discharge date: 17 Aug 1815 - Comments: Listed as a Negro in his service record; discharged at Elkton, MD, because of severe syphilis.

Harvey, William - Private - U.S. Light Artillery - Company: John Bell - Age: 40 - Height: 5' 11" - Eyes: Black - Hair: Black - Complexion: Black - Born: NH, Oxford County, Oxford - Trade: Farmer - Enlistment date: 15 Jul 1812 - Enlistment place: VT, Burlington - Enlistment period: War - Enlisted by whom: William Campbell - Discharge date: 31 Mar 1815 - Pension: Old War 10209; private in Second Lieutenant Nathaniel Dana's Detachment, U.S. Light Artillery - Comments: Possible Black, not verified; discharged at Boston, MA.

Hatford, Samuel - Laborer - Corps of Artificers - Company: Alexander Paris - Comments: Listed as "Black" in *A List of Pensioners of the War of 1812*.

Hawkins, George - Recruit - 26th U.S. Infantry - Recruiting Detachment: William Bezeau - Age: 21 - Height: 5' 9" - Eyes: Black - Hair: Curly - Complexion: Yellow - Born: New Jersey - Trade: Laborer - Enlistment date: 13 Oct 1814 - Enlistment place: Philadelphia - Enlistment period: War - Enlisted by whom: William Bezeau - Comments: Possible Black, not verified; deserted on 1 Nov 1814.

Hazard, Daniel - Recruit - 26th U.S. Infantry - Recruiting Detachment: William Bezeau - Age: 27 - Height: 5' 9" - Eyes: Black - Hair: Curly - Complexion: Brown - Born: Pennsylvania - Trade: Farmer - Enlistment date: 5 Feb 1815 - Enlistment place: Philadelphia - Enlistment period: War - Enlisted by whom: William Bezeau - Comments: Listed on the Descriptive Roll of Colored Men, 1 Apr 1815.

Hill, Isaac - Recruit - 26th U.S. Infantry - Recruiting Detachment: William Bezeau - Age: 25 - Height: 5' 7" - Eyes: Black - Hair: Black - Complexion: Chestnut - Born: PA, Philadelphia - Trade: Laborer - Enlistment date: 7 Oct 1814 - Enlistment place: Philadelphia - Enlistment period: War - Enlisted by whom: William Bezeau - Discharge date: 23 Mar 1815 - Bounty: BLW 7192-160-12 (Illinois) - Comments: Listed as "Col'd" in his service record; discharged at Philadelphia, PA.

Hill, Levin - Recruit - 26th U.S. Infantry - Recruiting Detachment: William Bezeau - Age: 22 - Height: 5' 7 1/2" - Eyes: Black - Hair: Curly - Complexion: Black - Born: Pennsylvania - Trade: Laborer - Enlistment date: 10 Oct 1814 - Enlistment place: Philadelphia - Enlistment period: 5 Yrs - Enlisted by whom: William Bezeau - Comments: Listed as "Col'd" in his service record; deserted on 12 Jan 1815.

Holfer, Michael - Recruit - 26th U.S. Infantry - Recruiting Detachment: William Bezeau - Age: 19 - Height: 5' 6" - Eyes: Blue - Hair: Curly - Complexion: Black - Born: Pennsylvania - Trade: Shoemaker - Enlistment date: 24 Jan 1815 - Enlistment place: Philadelphia - Enlistment period: War - Enlisted by whom: William Bezeau - Comments: Listed on the Descriptive Roll of Colored Men, 1 Apr 1815.

Horsey, James - Recruit - 26th U.S. Infantry - Recruiting Detachment: William Bezeau - Age: 28 - Height: 5' 9" - Eyes: Black - Hair: Black - Complexion: Black - Born: New Jersey - Trade: Laborer - Enlistment date: 17 Oct 1814 - Enlistment place: Philadelphia - Enlistment period: War - Enlisted by whom: William Bezeau - Comments: Listed as "Col'd" in his service record.

Howard, Perry - Recruit (Corporal) - 26th U.S. Infantry - Recruiting Detachment: William Bezeau - Enlistment date: 27 Sep 1814 - Enlistment period: War - Enlisted by whom: James Whelpley - Comments: Listed as "Col'd" in his service record.

Howell, Jeffry - Recruit - 26th U.S. Infantry - Recruiting Detachment: William Bezeau - Age: 24 - Height: 5' 7" - Eyes: Black - Hair: Curly - Complexion: Black - Born: PA, Philadelphia - Trade: Laborer - Enlistment date: 14 Oct 1814 - Enlistment place: Philadelphia - Enlistment period: War - Enlisted by whom: William Bezeau - Comments: Listed as "Colored" in his service record; deserted on 18 Oct 1814.

Hubbert, Jeremiah (Hubbard) - Recruit - 26th U.S. Infantry - Recruiting Detachment: William Bezeau - Age: 25 - Height: 5' 8" - Eyes: Black - Hair: Curly or Black - Complexion: Brown or Black - Born: Delaware - Enlistment date: 5 Jan 1815 - Enlistment place: Philadelphia - Enlistment period: War - Enlisted by whom: William Bezeau - Discharge date: 20 May 1815 - Comments: Listed as a Negro in his service record; discharged at Philadelphia, PA.

Huggs, Phil - Recruit - 26th U.S. Infantry - Recruiting Detachment: William Bezeau - Age: 25 - Height: 5' 6" - Eyes: Black - Hair: Black - Complexion: Black - Born: New Jersey - Trade: Farmer - Enlistment date: 14 Oct 1814 - Enlistment place: Philadelphia - Enlistment period: War - Enlisted by whom: William Bezeau - Discharge date: 23 Mar 1815 - Comments: Listed as "Col'd" in his service record; discharged at Philadelphia, PA.

Hullings, Isaac (Hulling) - Recruit - 26th U.S. Infantry - Recruiting Detachment: William Bezeau - Age: 29 - Height: 6' - Eyes: Black - Hair: Curly - Complexion: Black - Born: Pennsylvania - Trade: Farmer - Enlistment date: 16 Jan 1815 - Enlistment place: Philadelphia - Enlistment period: War - Enlisted by whom: William Bezeau - Comments: Listed on the Descriptive Roll of Colored Men, 1 Apr 1815; deserted on 27 Jan 1815.

Hunter, William - Recruit - 26th U.S. Infantry - Recruiting Detachment: William Bezeau - Age: 34 - Height: 6' - Eyes: Black - Hair: Curly - Complexion: Black - Born: Delaware - Trade: Seaman - Enlistment date: 12 Feb 1815 - Enlistment place: Philadelphia - Enlistment period: War - Enlisted by whom: William Bezeau - Comments: Listed on the Descriptive Roll of Colored Men, 1 Apr 1815.

Hupler, Daniel - Recruit - 26th U.S. Infantry - Recruiting Detachment: William Bezeau - Age: 27 - Height: 6' - Eyes: Black - Hair: Curly - Complexion: Black - Born: Pennsylvania - Trade: Farmer - Enlistment date: 9 Jun 1815 - Enlistment place: Philadelphia - Enlistment period: War - Enlisted by whom: William Bezeau - Comments: Listed on the Descriptive Roll of Colored Men, 1 Apr 1815; deserted on 18 Jan 1815.

Jackson, Gabriel - Private - U.S. Light Artillery - Company: John Bell - Age: 22 - Height: 5' 5 1/2" - Eyes: Black - Hair: Black - Complexion: Black - Born: DE, Wilmington - Trade: Mariner - Enlistment date: 10 Oct 1814 - Enlistment place: MA, Boston - Enlistment period: War - Enlisted by whom: William Campbell - Discharge date: 31 Mar 1815 - Bounty: BLW 6808-160-12 (Missouri) - Comments: Possible Black, not verified; discharged at Boston, MA.

Jackson, George - Private - U.S. Light Artillery - Company: John McIntosh - Age: 24 - Height: 5' 7 3/4" - Eyes: Black - Hair: Black - Complexion: Black - Born: NY, New York - Trade: Mariner - Enlistment date: 13 Jul 1814 - Enlistment place: MA, Boston - Enlistment period: War - Enlisted by whom: Samuel Washburn - Discharge date: 31 May 1815 - Comments: Possible Black, not verified; discharged at Plattsburgh, NY.

Jackson, George - Recruit - 26th U.S. Infantry - Recruiting Detachment: William Bezeau - Age: 30 - Height: 5' 5" - Eyes: Black - Hair: Curly - Complexion: Black - Born: Maryland - Trade: Farmer - Enlistment date: 25 Dec 1814 - Enlistment place: Philadelphia - Enlistment period: 5 Yrs - Enlisted by whom: William Bezeau - Comments: Listed as "Col'd" in his service record.

Jackson, James Z. - Recruit - 26th U.S. Infantry - Recruiting Detachment: William Bezeau - Age: 22 - Height: 5' 6" - Eyes: Black - Hair: Curly - Complexion: Black - Born: Delaware - Trade: Sailor - Enlistment date: 12 Feb 1815 - Enlistment place: Philadelphia - Enlistment period: War - Enlisted by whom: William Bezeau - Discharge date: 20 Mar 1815 - Bounty: BLW 1042-320-12 (Arkansas) - Comments: Listed on the Descriptive Roll of Colored Men, 1 Apr 1815; discharged at Philadelphia, PA.

Jackson, John - Recruit - 26th U.S. Infantry - Recruiting Detachment: William Bezeau - Age: 22 - Height: 5' 6" - Eyes: Black - Hair: Curly - Complexion: Brown - Born: Pennsylvania - Trade:

Laborer - Enlistment date: 26 Dec 1814 - Enlistment place: Philadelphia - Enlistment period: War - Enlisted by whom: William Bezeau - Discharge date: 20 Mar 1815 - Bounty: BLW 536-320-12 (Illinois) - Comments: Possible Black, not verified; discharged at Philadelphia, PA.

Jackson, John - Private - U.S. Light Artillery - Company: John McIntosh - Age: 21 – Height: 5' 6" - Eyes: Black - Hair: Black - Complexion: Black - Born: LA, New Orleans - Trade: Mariner - Enlistment date: 15 Jul 1814 - Enlistment place: MA, Boston - Enlistment period: War - Enlisted by whom: Samuel Washburn - Discharge date: 31 May 1815 - Comments: Possible Black, not verified; discharged at Plattsburg, NY.

Jackson, John Jr. - Recruit - 26th U.S. Infantry - Recruiting Detachment: William Bezeau - Age: 25 - Height: 5' 11" - Eyes: Black - Hair: Curly - Complexion: Black - Born: Connecticut - Trade: Farmer - Enlistment date: 15 Nov 1814 - Enlistment place: Philadelphia - Enlistment period: War - Enlisted by whom: William Bezeau - Comments: Listed as "Col'd" in his service record; deserted on 5 Dec 1814.

Jackson, William - Recruit - 26th U.S. Infantry - Recruiting Detachment: William Bezeau - Age: 25 - Height: 5' 5" - Eyes: Black - Hair: Curly - Complexion: Black - Born: Georgia - Enlistment date: 24 Jan 1815 - Enlistment place: Philadelphia - Enlistment period: War - Enlisted by whom: William Bezeau - Comments: Listed on the Descriptive Roll of Colored Men, 1 Apr 1815.

James, David - Recruit - 26th U.S. Infantry - Recruiting Detachment: William Bezeau - Age: 21 - Height: 5' 9" - Eyes: Grey - Hair: Curly - Complexion: Yellow - Born: Pennsylvania - Trade: Laborer - Enlistment date: 12 Feb 1815 - Enlistment place: Philadelphia - Enlistment period: 5 Yrs - Enlisted by whom: William Bezeau - Died: 2 Mar 1815 - Bounty: BLW 19232-160-12 Cancelled - Comments: Listed on the Descriptive Roll of Colored Men, 1 Apr 1815; land bounty to Sarah James, sister & only heir at law of David James, deceased; land bounty annulled, not entitled to land.

James, George - Recruit - 26th U.S. Infantry - Recruiting Detachment: William Bezeau - Age: 21 - Height: 5' 8 1/2" - Eyes: Black - Hair: Black - Complexion: Sallow - Born: Pennsylvania - Trade: Farmer - Enlistment date: 13 Sep 1814 - Enlistment place: Philadelphia - Enlistment period: War - Enlisted by whom: William Bezeau - Comments: Listed as "Col'd" in his service record; deserted on 25 Sep 1814.

James, James Frederick - 10th U.S. Infantry - Age: 20 - Height: 5' 3 5/8" - Eyes: Man of color - Hair: Man of color - Complexion: Man of color - Born: NC, Bertie County - Trade: Wheelwright - Enlistment date: 17 Aug 1812 - Enlistment period: 5 Yrs - Comments: Listed as a "Man of Color" in his service record; certificate dated NC, 17 Aug 1812.

Jamison, Edward - Recruit - 26th U.S. Infantry - Recruiting Detachment: William Bezeau - Age: 22 - Height: 5' 4" - Eyes: Black - Hair: Curly - Complexion: Yellow - Born: Pennsylvania - Trade: Seaman - Enlistment date: 9 Feb 1815 - Enlistment place: Philadelphia - Enlistment period: War - Enlisted by whom: William Bezeau - Bounty: BLW 833-320-14 Cancelled; BLW 1040-320-14 - Comments: Listed on the Descriptive Roll of Colored Men, 1 Apr 1815; discharged under the general order of 4 Mar 1815; land bounty to Abraham Jamison, cousin & only heir at law of Edward Jamison, deceased; died before 21 Oct 1819; first bounty was cancelled because Jamison was entitled to 320 acres of land.

Jasper, Peter - Recruit - 26th U.S. Infantry - Recruiting Detachment: William Bezeau - Age: 18 - Height: 5' 5" - Eyes: Black - Hair: Curly - Complexion: Black - Born: Maryland - Trade: Laborer -

Enlistment date: 21 Jan 1815 - Enlistment place: Philadelphia - Enlistment period: War - Enlisted by whom: William Bezeau - Comments: Listed on the Descriptive Roll of Colored Men, 1 Apr 1815; deserted on 27 Jan 1815.

Jefferson, Isaac - Recruit - 26th U.S. Infantry - Recruiting Detachment: William Bezeau - Age: 28 - Height: 5' 6" - Eyes: Black - Hair: Curly - Complexion: Black - Born: Pennsylvania - Trade: Blacksmith - Enlistment date: 4 Dec 1814 - Enlistment place: Philadelphia - Enlistment period: War - Enlisted by whom: William Bezeau - Discharge date: 1 May 1815 - Comments: Listed as "Col'd" in his service record; discharged at Philadelphia, PA.

Jister, Alexander - Recruit - 26th U.S. Infantry - Recruiting Detachment: William Bezeau - Age: 28 - Height: 5' 4" - Eyes: Black - Hair: Curly - Complexion: Black - Born: Pennsylvania - Trade: Farmer - Enlistment date: 18 Jan 1815 - Enlistment place: Philadelphia - Enlistment period: War - Enlisted by whom: William Bezeau - Comments: Listed on the Descriptive Roll of Colored Men, 1 Apr 1815.

Johnson, Alexander J. - Recruit - 26th U.S. Infantry - Recruiting Detachment: William Bezeau - Age: 31 - Height: 5' 4" - Eyes: Black - Hair: Curly - Complexion: not stated - Born: Africa - Trade: Waiter - Enlistment date: 7 Dec 1814 - Enlistment place: Philadelphia - Enlistment period: War - Enlisted by whom: William Bezeau - Discharge date: 20 Mar 1815 - Bounty: BLW 5288-160-12 (Illinois) - Comments: Listed as "Colored" in his service record; discharged at Philadelphia, PA.

Johnson, Amos - Private - 24th U.S. Infantry - Company: Silas Stephens - Age: 33 - Height: 5' 7" - Eyes: Black or Dark - Hair: Dark - Complexion: Black or Dark - Born: VA, Mecklenburg County - Trade: Farmer - Enlistment date: 9 Dec 1814 - Enlistment period: 5 Yrs - Enlisted by whom: William Chilton - Comments: Listed as "Black" in his service record.

Johnson, Isaac - Recruit - 26th U.S. Infantry - Recruiting Detachment: William Bezeau - Age: 40 - Height: 5' 11" - Eyes: Hazel - Hair: Curly - Complexion: Yellow - Born: Delaware - Trade: Miller - Enlistment date: 25 Oct 1814 - Enlistment place: Philadelphia - Enlistment period: 5 Yrs - Enlisted by whom: William Bezeau - Comments: Listed as "Col'd" in his service record.

Johnson, Jacob - Recruit - 26th U.S. Infantry - Recruiting Detachment: William Bezeau - Age: 22 - Height: 5' 6" - Eyes: Black - Hair: Curly - Complexion: Black - Born: Pennsylvania - Trade: Farmer - Enlistment date: 21 Oct 1814 - Enlistment place: Philadelphia - Enlistment period: 5 Yrs - Enlisted by whom: William Bezeau - Discharge date: 20 May 1815 - Comments: Listed as "Col'd" in his service record.

Johnson, James - Recruit - 26th U.S. Infantry - Recruiting Detachment: William Bezeau - Age: 35 - Height: 5' 9" - Eyes: Black - Hair: Curly - Complexion: Yellow - Born: Maryland - Trade: Laborer - Enlistment date: 24 Nov 1814 - Enlistment place: Philadelphia - Enlistment period: War - Enlisted by whom: William Bezeau - Discharge date: 20 Mar 1815 - Bounty: BLW 10293-160-12 (Illinois) - Comments: Listed as "Col'd" in his service record; discharged at Philadelphia, PA.

Johnson, James F. - Recruit - 26th U.S. Infantry - Recruiting Detachment: William Bezeau - Age: 21 - Height: 5' 10" - Eyes: Black - Hair: Curly - Complexion: Brown - Born: Delaware - Trade: Sailor - Enlistment date: 12 Jan 1815 - Enlistment place: Philadelphia - Enlistment period: War - Enlisted by whom: William Bezeau - Discharge date: 20 May 1815 - Bounty: BLW 10294-160-12 (Illinois) - Comments: Listed on the Descriptive Roll of Colored Men, 1 Apr 1815; discharged at Philadelphia, PA.

Johnson, John - Private - U.S. Light Artillery - Company: George Morris - Age: 22 - Height: 5' 7 1/2" - Eyes: Black - Hair: Black - Complexion: Black - Born: CT, Litchfield County, Danbury - Trade: Laborer - Enlistment date: 22 Sep 1814 - Enlistment place: MA, Norfolk County, Dedham - Enlistment period: War - Enlisted by whom: William Campbell - Discharge date: 31 Mar 1815 - Comments: "Discharged on account of being a Negro and unfit to accompany American soldiers."

Johnson, Joseph T. - Recruit - 26th U.S. Infantry - Recruiting Detachment: William Bezeau - Age: 20 - Height: 5' 3" - Eyes: Black - Hair: Curly - Complexion: Black - Born: West Indies, Trinidad - Trade: Seaman - Enlistment date: 3 Feb 1815 - Enlistment place: Philadelphia - Enlistment period: War - Enlisted by whom: William Bezeau - Discharge date: 20 Mar 1815 - Bounty: BLW 311-320-14 (Illinois) - Comments: Listed on the Descriptive Roll of Colored Men, 1 Apr 1815; discharged at Philadelphia, PA.

Johnson, Richard - Private - U.S. Light Artillery - Company: George Morris - Age: 25 - Height: 5' 5" - Eyes: Dark - Hair: Black - Complexion: Black - Born: CT, Litchfield County, Danbury - Trade: Laborer - Enlistment date: 22 Sep 1814 - Enlistment place: MA, Norfolk County, Dedham - Enlistment period: 5 Yrs - Enlisted by whom: William Campbell - Discharge date: 31 Mar 1815 - Comments: "Discharged on account of being a Negro and not fit to accompany American soldiers."

Johnson, Samuel - Private - U.S. Light Artillery - Company: John Bell - Age: 30 - Height: 5' 8" - Eyes: Black - Hair: Black - Complexion: Black - Born: NY, Long Island - Trade: Mariner - Enlistment date: 29 Aug 1814 - Enlistment place: MA, Boston - Enlistment period: War - Enlisted by whom: Samuel Washburn - Discharge date: 31 Mar 1815 - Comments: Possible Black, not verified.

Johnson, Thomas - Recruit - 26th U.S. Infantry - Recruiting Detachment: William Bezeau - Age: 29 - Height: 5' 7" - Eyes: Black - Hair: Curly - Complexion: Brown - Born: DC, Washington - Trade: Laborer - Enlistment date: 13 Nov 1814 - Enlistment place: Philadelphia - Enlistment period: War - Enlisted by whom: William Bezeau - Discharge date: 25 Mar 1815 - Bounty: BLW 7195-160-12 (Illinois) - Comments: Listed as "Col'd" in his service record; discharged at Philadelphia, PA.

Johnson, William - Recruit - 26th U.S. Infantry - Recruiting Detachment: William Bezeau - Age: 28 - Height: 5' 4" - Eyes: Black - Hair: Black - Complexion: Black - Born: Rhode Island - Trade: Sail maker - Enlistment date: 3 Nov 1814 - Enlistment place: Philadelphia - Enlistment period: 5 Yrs - Enlisted by whom: William Bezeau - Discharge date: 1 May 1815 - Comments: Listed as "Colored" in his service record; discharged, not fit for military duty (medical reason).

Johnson, William - Private - 9th U.S. Infantry - Company: Otis Fisher - Age: 23 - Height: 5' 3 1/2" - Eyes: Black - Hair: Black - Complexion: Black - Born: MA, Quebec - Trade: Farmer - Enlistment date: 7 Dec 1814 - Enlistment place: MA, Pittsfield - Enlistment period: War - Enlisted by whom: Benoni Allen - Discharge date: 31 Mar 1815 - Comments: Possible Black, not verified; discharged at Pittsfield, MA.

Jones, Henry - Recruit - 26th U.S. Infantry - Recruiting Detachment: William Bezeau - Age: 20 - Height: 5' 6" - Eyes: Dark - Hair: Curly - Complexion: Black - Born: Delaware - Trade: Laborer - Enlistment date: 4 Dec 1814 - Enlistment place: Philadelphia - Enlistment period: War - Enlisted by whom: William Bezeau - Died: 9 Jan 1815 - Comments: Listed as "Col'd" in his service record; died in general hospital.

Jones, Ira - Private - 30th U.S. Infantry - Company: Sylvanus Danforth - Other regiment: U.S. Light Artillery - Age: 19 - Height: 5' 4" - Eyes: Black - Hair: Curly - Complexion: Black - Born: MA, Hampshire County, Worthington - Trade: Farmer - Enlistment date: 18 Dec 1814 - Enlistment place: Ballston - Enlistment period: 5 Yrs - Enlisted by whom: Gidcon Brownson - Comments: Listed as a Negro in his service record.

Jones, Isaac - Recruit - 26th U.S. Infantry - Recruiting Detachment: William Bezeau - Age: 44 - Height: 5' 9 1/2" - Eyes: Brown - Hair: Dark - Complexion: Dark - Born: Pennsylvania - Trade: Laborer - Enlistment date: 2 Dec 1814 - Enlistment place: Philadelphia - Enlistment period: 5 Yrs - Enlisted by whom: William Bezeau - Discharge date: 20 May 1815 - Comments: Listed as "Col'd" in his service record; discharged at Philadelphia, PA.

Jones, Solomon - Recruit - 26th U.S. Infantry - Recruiting Detachment: William Bezeau - Age: 40 - Height: 5' 8" - Eyes: Grey - Hair: Brown - Complexion: Dark - Born: Pennsylvania - Trade: Laborer - Enlistment date: 30 Aug 1814 - Enlistment place: Philadelphia - Enlistment period: 5 Yrs - Enlisted by whom: William Bezeau - Discharge date: 1 Jun 1815 - Comments: Listed as "Col'd" in his service record; discharged at Philadelphia, PA.

Joseph, Daniel - Recruit - 26th U.S. Infantry - Recruiting Detachment: William Bezeau - Age: 24 - Height: 5' 8" - Eyes: Black - Hair: Curly - Complexion: Black - Born: Delaware - Trade: Farmer - Enlistment date: 7 Feb 1815 - Enlistment place: Philadelphia - Enlistment period: War - Enlisted by whom: William Bezeau - Comments: Listed on the Descriptive Roll of Colored Men, 1 Apr 1815.

Joseph, John - Recruit - 26th U.S. Infantry - Recruiting Detachment: William Bezeau - Age: 23 - Height: 5' 5" - Eyes: Black - Hair: Curly - Complexion: Yellow - Born: Portugal, Lisbon - Trade: Mariner - Enlistment date: 25 Dec 1814 - Enlistment place: Philadelphia - Enlistment period: 5 Yrs - Enlisted by whom: William Bezeau - Discharge date: 20 May 1815 - Comments: Listed as "Col'd" and as a "Black man" in his service record; discharged at Philadelphia, PA.

Junior, Ceazer (or Caizer) - Recruit - 26th U.S. Infantry - Recruiting Detachment: William Bezeau - Age: 40 - Height: 5' 3" - Eyes: Black - Hair: Curly - Complexion: Black - Born: Pennsylvania - Trade: Laborer - Enlistment date: 29 Nov 1814 - Enlistment place: Philadelphia - Enlistment period: 5 Yrs - Enlisted by whom: William Bezeau - Discharge date: 20 May 1815 - Comments: Listed as "Col'd" in his service record.

Karmon, Peter (Kannon) - Recruit - 26th U.S. Infantry - Recruiting Detachment: William Bezeau - Age: 17 - Height: 5' 4" - Eyes: Black - Hair: Curly - Complexion: Brown - Born: Maryland - Trade: Farmer - Enlistment date: 16 Jan 1815 - Enlistment place: Philadelphia - Enlistment period: War - Enlisted by whom: William Bezeau - Comments: Listed on the Descriptive Roll of Colored Men, 1 Apr 1815.

Kellehorn, James (Kelliharn) - Private - U.S. Light Artillery - Company: John McIntosh - Age: 21 - Height: 5' 6" - Eyes: Black - Hair: Black - Complexion: Black - Born: MA, Suffolk County, Boston - Trade: Rope maker - Enlistment date: 19 Jul 1814 - Enlistment place: MA, Boston - Enlistment period: War - Enlisted by whom: Samuel Washburn - Discharge date: 31 May 1815 - Comments: Possible Black, not verified; discharged at Plattsburg, NY.

Kelley, Ezekiel - Recruit - 26th U.S. Infantry - Recruiting Detachment: William Bezeau - Age: 19 - Height: 5' 6" - Eyes: Black - Hair: Curly - Complexion: Black - Born: West Indies, Jamaica - Trade: Seaman - Enlistment date: 19 Jan 1815 - Enlistment place: Philadelphia - Enlistment period: War - Enlisted by whom: William Bezeau - Comments: Listed on the Descriptive Roll of Colored Men, 1 Apr 1815.

Kimbert, Moses - Recruit - 26th U.S. Infantry - Recruiting Detachment: William Bezeau - Age: 21 -

Height: 5' 8" - Eyes: Black - Hair: Curly - Complexion: Black - Born: New York - Trade: Farmer - Enlistment date: 2 Feb 1815 - Enlistment place: Philadelphia - Enlistment period: War - Enlisted by whom: William Bezeau - Comments: Listed on the Descriptive Roll of Colored Men, 1 Apr 1815.

Kingsley, Samuel - Recruit (Sergeant) - 26th U.S. Infantry - Recruiting Detachment: William Bezeau - Age: 21 - Height: 5' 6" - Born: MD, Harford County, Havre de Grace - Enlistment date: 8 Sep 1814 - Enlistment place: MD, Baltimore - Enlistment period: War - Enlisted by whom: James Whelpley - Comments: Listed as "Col'd" in his service record.

Kirkpatrick, Ezekiel - Private - 39th U.S. Infantry - Company: George Hallum - Enlistment date: 15 Jan 1815 - Discharge date: 14 Jan 1815 - Comments: Reported as a negro on the regimental muster rolls completed after the Battle of Horseshoe Bend.

Kisling, James - Recruit - 26th U.S. Infantry - Recruiting Detachment: William Bezeau - Age: 27 - Height: 5' 6" - Eyes: Black - Hair: Curly - Complexion: Brown - Born: Maryland - Trade: Farmer - Enlistment date: 19 Jan 1815 - Enlistment place: Philadelphia - Enlistment period: War - Enlisted by whom: William Bezeau - Comments: Listed on the Descriptive Roll of Colored Men, 1 Apr 1815.

Knight, John - Recruit - 26th U.S. Infantry - Recruiting Detachment: William Bezeau - Age: 21 - Height: 5' 6 1/2" - Eyes: Black - Hair: Curly - Complexion: Yellow - Born: Pennsylvania - Trade: Laborer - Enlistment date: 24 Nov 1814 - Enlistment place: Philadelphia - Enlistment period: War - Enlisted by whom: William Bezeau - Comments: Listed as "Col'd" in his service record.

La Cruse – see Francis L. Cruse

Ladine, Noah - Recruit - 26th U.S. Infantry - Recruiting Detachment: William Bezeau - Age: 28 - Height: 5' 8" - Eyes: Black - Hair: Black - Complexion: Black - Born: Delaware - Trade: Miller - Enlistment date: 7 Oct 1814 - Enlistment place: Philadelphia - Enlistment period: 5 Yrs - Enlisted by whom: William Bezeau - Discharge date: 24 May 1815 - Comments: Listed as "Colored" in his service record; discharged at Philadelphia, PA; he has two service records recorded. The second service record is listed under Noah Lowdine, and it indicates that he is a "Blackman".

Ladman, Charles - Recruit - 26th U.S. Infantry - Recruiting Detachment: William Bezeau - Age: 31 - Height: 5' 8" - Eyes: Black - Hair: Curly - Complexion: Brown - Born: Pennsylvania - Trade: Farmer - Enlistment date: 13 Jan 1815 - Enlistment place: Philadelphia - Enlistment period: War - Enlisted by whom: William Bezeau - Comments: Probably Black, not verified; not listed on the Descriptive Roll of Colored Men, 1 Apr 1815, but on the Descriptive Roll for 1 Apr 1815 (for Whites).

Lambert, George - Private - 26th US Infantry - Company: William Bezeau - Enlistment date: 5 Jan 1815 - Enlistment period: War - Enlisted by whom: William Bezeau - Comments: Colored man.

Lambert, Samuel - Recruit - 26th U.S. Infantry - Recruiting Detachment: William Bezeau - Age: 24 - Height: 5' 10" - Eyes: Black - Hair: Curly - Complexion: Brown - Born: Pennsylvania - Trade: Laborer - Enlistment date: 19 Jan 1815 - Enlistment place: Philadelphia - Enlistment period: War - Enlisted by whom: William Bezeau - Comments: Listed on the Descriptive Roll of Colored Men, 1 Apr 1815.

Lang, William - Recruit - 26th U.S. Infantry - Recruiting Detachment: William Bezeau - Age: 22 - Height: 5' 6" - Eyes: Black - Hair: Curly - Complexion: Dark - Born: Pennsylvania - Trade: Farmer

- Enlistment date: 11 Nov 1814 - Enlistment place: Philadelphia - Enlistment period: War - Enlisted by whom: William Bezeau - Discharge date: 23 Mar 1815 - Bounty: BLW 7195-160-12 (Illinois) - Comments: Listed as "Colored" in his service record; discharged at Philadelphia, PA.

Larkin, Manuel A.- Private - 3rd U.S. Infantry - Company: Robert Moore - Age: 29 - Height: 5' 6" - Eyes: Black - Hair: Black - Complexion: Black - Trade: Laborer - Enlistment date: 3 May 1814 - Enlistment place: AL, Fort Claiborne - Enlistment period: 5 Yrs - Enlisted by whom: Robert Moore - Comments: Possible Black, not verified; deserted on 9 May 1815; also listed as Manuel or Manwell Alarkin.

Larkman, George - Recruit - 26th U.S. Infantry - Recruiting Detachment: William Bezeau - Age: 21 - Height: 5' 5" - Eyes: Black - Hair: Curly - Complexion: Black - Born: Pennsylvania - Trade: Laborer - Enlistment date: 7 Jan 1815 - Enlistment place: Philadelphia - Enlistment period: War - Enlisted by whom: William Bezeau - Comments: Listed on the Descriptive Roll of Colored Men, 1 Apr 1815.

Lawrence, Nathaniel - Laborer - Corps of Artificers - Company: Alexander Paris - Comments: Listed as "Black" in *A List of Pensioners of the War of 1812*.

Lawrence, Peter - Private - U.S. Light Artillery - Company: George Morris - Age: 23 - Height: 5' 8" - Eyes: Black - Hair: Black - Complexion: Black - Born: VA, Petersburg - Trade: Mariner - Enlistment date: 7 Oct 1814 - Enlistment place: MA, Boston - Enlistment period: War - Enlisted by whom: William Campbell - Discharge date: 31 Mar 1815 - Bounty: BLW 13175-160-12 (Illinois) - Comments: Possible Black, not verified; discharged at Boston, MA.

Leadway – see Benjamin Lodway

Lee, Andrew - Private - U.S. Light Artillery - Company: John Bell - Age: 21 - Height: 5' 7 1/2" - Eyes: Black - Hair: Black - Complexion: Black - Born: MA, Hampshire County, Northampton - Trade: Farmer - Enlistment date: 23 Sep 1814 - Enlistment place: MA, Boston - Enlistment period: War - Enlisted by whom: William Campbell - Discharge date: 31 Mar 1815 - Comments: Possible Black, not verified.

Lee, Jeremiah - Recruit - 26th U.S. Infantry - Recruiting Detachment: William Bezeau - Age: 23 - Height: 5' 7" - Eyes: Black - Hair: Curly or Black - Complexion: Black - Born: Delaware - Trade: Laborer - Enlistment date: 1 Jan 1815 - Enlistment place: Philadelphia - Enlistment period: 5 Yrs - Enlisted by whom: William Bezeau - Discharge date: 20 May 1815 - Comments: Listed as "Colored" and as a "Blackman" in his service record; discharged at Philadelphia, PA.

Lee, Thomas - Recruit - 26th U.S. Infantry - Recruiting Detachment: William Bezeau - Age: 19 - Height: 5' 6" - Eyes: Hazel - Hair: Curly - Complexion: Black - Born: New Jersey - Trade: Laborer - Enlistment date: 6 Jan 1815 - Enlistment place: Philadelphia - Enlistment period: War - Enlisted by whom: William Bezeau - Discharge date: 20 Mar 1815 - Bounty: BLW 741-320-12 (Illinois) - Comments: Listed on the Descriptive Roll of Colored Men, 1 Apr 1815.

Leubert, George - Recruit - 26th U.S. Infantry - Recruiting Detachment: William Bezeau - Age: 31 - Height: 5' 5" - Eyes: Black - Hair: Curly - Complexion: Black - Born: Pennsylvania - Trade: Laborer - Enlistment date: 5 Jan 1815 - Enlistment place: Philadelphia - Enlistment period: War - Enlisted by whom: William Bezeau - Comments: Listed on the Descriptive Roll of Colored Men, 1 Apr 1815.

Lewis, Benjamin - Recruit - 26th U.S. Infantry - Recruiting Detachment: William Bezeau - Age: 29 - Height: 5' 6" - Eyes: Black - Hair: Curly - Complexion: Brown - Born: Pennsylvania - Trade: Laborer - Enlistment date: 2 Nov 1814 - Enlistment place: Philadelphia - Enlistment period: War - Enlisted by whom: William Bezeau - Comments: Listed as "Col'd" in his service record.

Lewis, Prince - Recruit - 26th U.S. Infantry - Recruiting Detachment: William Bezeau - Age: 23 - Height: 5' 8 1/2" - Eyes: Black - Hair: Curly - Complexion: Mulatto - Born: New Jersey - Trade: Farmer - Enlistment date: 30 Dec 1814 - Enlistment place: Philadelphia - Enlistment period: War - Enlisted by whom: William Bezeau - Discharge date: 31 Mar 1815 - Bounty: BLW 743-320-12 (Illinois) - Comments: Listed as "Col'd" in his service record.

Lewis, Robert - Private - U.S. Light Artillery - Company: George Morris - Age: 22 - Height: 5' 2 7/8" - Eyes: Black - Hair: Black - Complexion: Black - Born: VA, Chesterfield County, Chester - Trade: Mariner - Enlistment date: 3 Aug 1814 - Enlistment place: MA, Salem or Boston - Enlistment period: War - Enlisted by whom: Samuel Washburn - Discharge date: 31 Mar 1815 - Comments: Possible Black, not verified; discharged at Boston, MA.

Lillman, Joseph - Recruit - 26th U.S. Infantry - Recruiting Detachment: William Bezeau - Enlistment date: 20 Sep 1814 - Enlistment place: Philadelphia - Enlistment period: War - Enlisted by whom: William Bezeau - Comments: Listed as "Col'd" in his service record.

Limmon, Charles - Recruit - 26th U.S. Infantry - Recruiting Detachment: William Bezeau - Age: 25 - Height: 5' 7" - Eyes: Black - Hair: Curly - Complexion: Black - Born: Pennsylvania - Trade: Laborer - Enlistment date: 15 Jan 1815 - Enlistment place: Philadelphia - Enlistment period: War - Enlisted by whom: William Bezeau - Comments: Listed on the Descriptive Roll of Colored Men, 1 Apr 1815.

Linds, Samuel - Private - 11th U.S. Infantry - Company: Benjamin Smead - Age: 20 - Height: 5' 10" - Eyes: Black - Hair: Black - Complexion: Black color - Trade: Husbandman - Enlistment date: 1 Mar 1813 - Enlistment period: 5 Yrs - Enlisted by whom: Benjamin Smead - Discharge date: 19 Sep 1815 - Comments: Possible Black, not verified; discharged at Fort Columbus, Governor's Island, NY, for disability.

Lines – see William Lynes

Liston, John - Recruit - 26th U.S. Infantry - Recruiting Detachment: William Bezeau - Age: 24 - Height: 5' 5" - Eyes: Black - Hair: Curly - Complexion: Dark - Born: PA, Philadelphia - Trade: Farmer - Enlistment date: 21 Nov 1814 - Enlistment place: Philadelphia - Enlistment period: 5 Yrs - Enlisted by whom: William Bezeau - Discharge date: 7 May 1815 - Comments: Listed as "Col'd" in his service record; discharged at Philadelphia, PA.

Little, Lymus W. - Private - U.S. Light Artillery - Age: 35 - Height: 5' 4 1/2" - Eyes: Black - Hair: Black - Complexion: Black - Born: CT, Windham County, Painfield - Trade: Mariner - Enlistment date: 10 Oct 1814 - Enlistment place: MA, Boston - Enlistment period: War - Enlisted by whom: William Campbell - Discharge date: 31 Mar 1815 - Comments: Possible Black, not verified.

Littleton, Samuel - Private - 31st U.S. Infantry - Company: Rufus Stewart - Age: 39 - Height: 5' 7" - Eyes: Black - Hair: Black - Complexion: Black - Born: MA, Charlestown - Trade: Farmer - Enlistment date: 26 Apr 1814 - Enlistment place: VT, Montpelier - Enlistment period: War - Enlisted by whom: Harvey Gillman - Discharge date: 3 Jun 1815 - Comments: Possible Black, not verified; discharged at Plattsburg, NY.

Lodway, Benjamin - Recruit - 26th U.S. Infantry - Recruiting Detachment: William Bezeau - Age: 27 - Height: 5' 8" - Eyes: Black - Hair: Curly - Complexion: Black - Born: Pennsylvania - Trade: Laborer - Enlistment date: 23 Jan 1815 - Enlistment place: Philadelphia - Enlistment period: War - Enlisted by whom: William Bezeau - Comments: Listed on the Descriptive Roll of Colored Men, 1 Apr 1815.

Looks, Samuel - Musician - 2nd U.S. Rifles - Company: Batteal Harrison - Age: 20 - Height: 5' 5" - Eyes: Black - Hair: Black - Complexion: Black - Born: Canada, Quebec - Trade: Musician - Enlistment date: 4 Jul 1814 - Enlistment place: OH, Ross County, Chillicothe - Enlistment period: 5 Yrs - Enlisted by whom: Batteal Harrison - Comments: Possible Black, not verified.

Lowdine – see Noah Ladine

Lower, Joseph - Recruit - 26th U.S. Infantry - Recruiting Detachment: William Bezeau - Age: 28 - Height: 5' 6" - Eyes: Black - Hair: Curly - Complexion: Dark - Born: Pennsylvania - Trade: Laborer - Enlistment date: 6 Feb 1815 - Enlistment place: Philadelphia - Enlistment period: War - Enlisted by whom: William Bezeau - Comments: Listed as "Col'd" in his service record.

Loyd, Samuel - Recruit - 26th U.S. Infantry - Recruiting Detachment: William Bezeau - Age: 36 - Height: 5' 9 1/2" - Eyes: Black - Hair: Curly - Complexion: Brown - Born: Pennsylvania - Trade: Farmer - Enlistment date: 14 Jan 1815 - Enlistment place: Philadelphia - Enlistment period: War - Enlisted by whom: William Bezeau - Comments: Listed on the Descriptive Roll of Colored Men, 1 Apr 1815.

Lubby, Thomas - Recruit - 26th U.S. Infantry - Recruiting Detachment: William Bezeau - Age: 30 - Height: 5' 11" - Eyes: Black - Hair: Curly - Complexion: Black - Born: Delaware - Trade: Laborer - Enlistment date: 9 Feb 1815 - Enlistment place: Philadelphia - Enlistment period: War - Enlisted by whom: William Bezeau - Comments: Listed on the Descriptive Roll of Colored Men, 1 Apr 1815.

Luff, Henry - Recruit - 26th U.S. Infantry - Recruiting Detachment: William Bezeau - Age: 27 - Height: 5' 9" - Eyes: Dark - Hair: Black - Complexion: Brown - Born: Pennsylvania - Trade: Sailor - Enlistment date: 27 May 1814 - Enlistment place: Philadelphia - Enlistment period: 5 Yrs - Enlisted by whom: William Bezeau - Comments: Listed as "Col'd" in his service record.

Luke, Seth - Recruit - 26th U.S. Infantry - Recruiting Detachment: William Bezeau - Age: 26 - Height: 6' - Eyes: Black - Hair: Curly - Complexion: Black - Born: Pennsylvania - Trade: Laborer - Enlistment date: 27 Jan 1815 - Enlistment place: Philadelphia - Enlistment period: War - Enlisted by whom: William Bezeau - Comments: Listed on the Descriptive Roll of Colored Men, 1 Apr 1815.

Lynes, William - Private - 30th U.S. Infantry - Company: Gideon Spencer - Age: 21 - Height: 5' 3' - Eyes: Black - Hair: Curly - Complexion: Black - Born: MA, Berkshire County, Williamston - Trade: Farmer - Enlistment date: 2 Apr 1814 - Enlistment place: VT, Burlington - Enlistment period: 5 Yrs - Enlisted by whom: William Barney - Comments: Possible Black, not verified; on duty with the U.S. Navy after 6 Sep 1814; deserted at Burlington on 30 Apr 1815.

Madber, Thomas - Recruit - 26th U.S. Infantry - Recruiting Detachment: William Bezeau - Age: 24 - Height: 5' 11" - Eyes: Black - Hair: Curly - Complexion: Black - Born: Pennsylvania - Trade: Farmer - Enlistment date: 23 Jan 1815 - Enlistment place: Philadelphia - Enlistment period: War - Enlisted by whom: William Bezeau - Comments: Listed on the Descriptive Roll of Colored Men, 1

Apr 1815.

Maddon – see Joseph Midden

Maddy, Levin - Recruit - 26th U.S. Infantry - Recruiting Detachment: William Bezeau - Age: 30 - Height: 5' 6" - Eyes: Black - Hair: Black - Complexion: Black - Born: Maryland - Trade: Laborer - Enlistment date: 9 Oct 1814 - Enlistment place: Philadelphia - Enlistment period: War - Enlisted by whom: William Bezeau - Discharge date: 20 Mar 1815 - Bounty: BLW 10295-160-12 (Illinois) - Comments: Listed as "Col'd" in his service record; discharged at Philadelphia, PA.

Manner, James - Recruit - 26th U.S. Infantry - Recruiting Detachment: William Bezeau - Age: 22 - Height: 5' 4" - Eyes: Black - Hair: Curly - Complexion: Black - Born: New Jersey - Trade: Farmer - Enlistment date: 8 Feb 1815 - Enlistment place: Philadelphia - Enlistment period: War - Enlisted by whom: William Bezeau - Comments: Listed on the Descriptive Roll of Colored Men, 1 Apr 1815.

Marcus, Richard - Recruit - 26th U.S. Infantry - Recruiting Detachment: William Bezeau - Age: 20 - Height: 5' 6 1/2" - Eyes: Black - Hair: Curly - Complexion: Brown - Born: Pennsylvania - Trade: Laborer - Enlistment date: 18 Dec 1814 - Enlistment place: Philadelphia - Enlistment period: War - Enlisted by whom: William Bezeau - Comments: Listed as "Col'd" in his service record.

Marshall, Samuel W. - Laborer - Corps of Artificers - Company: Alexander Paris - Comments: Listed as "Black" in *A List of Pensioners of the War of 1812*.

Masters, William - Private - U.S. Light Artillery - Company: George Melven - Age: 17 - Height: 5' 3 1/2" - Eyes: Black - Hair: Black - Complexion: Black - Born: NH, Merrimack County, Andover - Trade: Blacksmith - Enlistment date: 5 Feb 1813 - Enlistment place: NH, Windsor - Enlistment period: 5 Yrs - Enlisted by whom: James Cobb - Discharge date: 28 Aug 1815 - Comments: Possible Black, not verified; discharged by civil authority.

Mathias, Charles (Mathews) - Recruit - 26th U.S. Infantry - Recruiting Detachment: William Bezeau - Age: 26 - Height: 5' - Eyes: Brown - Hair: Black - Complexion: Chestnut - Born: Maryland - Trade: Shoemaker - Enlistment date: 8 Oct 1814 - Enlistment place: Philadelphia - Enlistment period: 5 Yrs - Enlisted by whom: William Bezeau - Discharge date: 16 May 1815 - Comments: Listed as "Colored" in his service record; discharged at Philadelphia, PA.

Matthews, Andrew - Recruit - 26th U.S. Infantry - Recruiting Detachment: William Bezeau - Age: 30 - Height: 5' 5" - Eyes: Black - Hair: Curly - Complexion: Black - Born: Ohio - Trade: Seaman - Enlistment date: 9 Feb 1815 - Enlistment place: Philadelphia - Enlistment period: War - Enlisted by whom: William Bezeau - Comments: Listed on the Descriptive Roll of Colored Men, 1 Apr 1815.

Maxwell, Samuel - Recruit - 26th U.S. Infantry - Recruiting Detachment: William Bezeau - Age: 19 - Height: 5' 6" - Eyes: Black - Hair: Curly - Complexion: Black - Born: Pennsylvania - Trade: Farmer - Enlistment date: 27 Jan 1815 - Enlistment place: Philadelphia - Enlistment period: War - Enlisted by whom: William Bezeau - Comments: Listed on the Descriptive Roll of Colored Men, 1 Apr 1815.

McIntire, Solomon - Recruit - 26th U.S. Infantry - Recruiting Detachment: William Bezeau - Age: 23 - Height: 5' 8" - Eyes: Black - Hair: Curly - Complexion: Yellow - Born: New Jersey - Trade: Farmer - Enlistment date: 10 Nov 1814 - Enlistment place: Philadelphia - Enlistment period: War - Enlisted by whom: William Bezeau - Discharge date: 23 Mar 1815 - Bounty: BLW 7198-160-12

(Illinois) - Comments: Listed as "Col'd" in his service record; discharged at Philadelphia, PA.

McMullen, Cyrus - Recruit - 26th U.S. Infantry - Recruiting Detachment: William Bezeau - Age: 29 - Height: 5' 8" - Eyes: Black - Hair: Curly - Complexion: Black - Born: Louisiana - Trade: Seaman - Enlistment date: 3 Feb 1815 - Enlistment place: Philadelphia - Enlistment period: War - Enlisted by whom: William Bezeau - Comments: Listed on the Descriptive Roll of Colored Men, 1 Apr 1815.

McMutrie, Samuel - Recruit - 26th U.S. Infantry - Recruiting Detachment: William Bezeau - Age: 36 - Height: 5' 10" - Eyes: Black - Hair: Black - Complexion: Brown - Born: New Jersey - Trade: Laborer - Enlistment date: 23 Oct 1814 - Enlistment place: Philadelphia - Enlistment period: War - Enlisted by whom: William Bezeau - Discharge date: 23 Mar 1815 - Bounty: BLW 7200-160-12 (Illinois) - Comments: Listed as "Col'd" in his service record; discharged at Philadelphia, PA.

Meirs, Shadrack - Recruit - 26th U.S. Infantry - Recruiting Detachment: William Bezeau - Age: 15 - Height: 4' 6" - Eyes: Black - Hair: Black - Complexion: Black - Born: Pennsylvania - Trade: Laborer - Enlistment date: 11 Oct 1814 - Enlistment place: Philadelphia - Enlistment period: War - Enlisted by whom: William Bezeau - Comments: Listed as "Col'd" in his service record; discharged for being a minor.

Meldrom, James - Recruit - 26th U.S. Infantry - Recruiting Detachment: William Bezeau - Age: 35 - Height: 5' 10 1/2" - Eyes: Black - Hair: Black - Complexion: Yellow - Born: New Jersey - Trade: Farmer - Enlistment date: 9 Oct 1814 - Enlistment place: Philadelphia - Enlistment period: 5 Yrs - Enlisted by whom: William Bezeau - Discharge date: 25 May 1815 - Comments: Listed as "Col'd" in his service record; discharged at Philadelphia, PA.

Melona, Cornelius - Laborer - Corps of Artificers - Company: Alexander Paris - Comments: Listed as "Black" in *A List of Pensioners of the War of 1812*.

Midden, Joseph - Recruit - 26th U.S. Infantry - Recruiting Detachment: William Bezeau - Age: 21 - Height: 6' - Eyes: Black - Hair: Curly - Complexion: Black - Born: Pennsylvania - Trade: Seaman - Enlistment date: 7 Feb 1815 - Enlistment place: Philadelphia - Enlistment period: War - Enlisted by whom: William Bezeau - Comments: Listed on the Descriptive Roll of Colored Men, 1 Apr 1815.

Milleman, Prince - Private - 30th US Infantry - Company: Sylvanus Danforth - Other regiment: US Light Artillery - Age: 36 - Height: 5' 8" - Eyes: Black - Hair: Curly - Complexion: Negro - Born: RI, Newport - Trade: Farmer - Enlistment date: 27 Oct 1814 - Enlistment place: VT, Middlebury - Enlistment period: 5 Yrs - Enlisted by whom: Simeon Wright - Comments: Transferred to the U.S. Light Artillery, Captain Nathan Towson's Company; deserted at Fort Warren, VT, on 2 Aug 1816.

Miller, Abraham - Private - 16th U.S. Infantry - Company: Robert Gray - Age: 44 - Height: 5' 8" - Eyes: Black - Hair: Black - Complexion: Black - Born: Pennsylvania - Trade: Weaver - Enlistment date: 15 Sep 1812 - Enlistment place: PA, Marietta - Enlistment period: 5 Yrs - Enlisted by whom: Terah. Jones - Comments: Possible Black, not verified.

Miller, William - Recruit - 26th U.S. Infantry - Recruiting Detachment: William Bezeau - Age: 22 - Height: 5' - Eyes: Black - Hair: Black - Complexion: Black or Dark - Born: Pennsylvania - Trade: Farmer - Enlistment date: 10 Sep 1814 - Enlistment place: Philadelphia - Enlistment period: 5 Yrs - Enlisted by whom: William Bezeau - Discharge date: 22 May 1815 – Comments: Listed as a "Blackman" in his service record; discharged at Philadelphia, PA.

Millikin, James - Laborer - Corps of Artificers - Company: Alexander Paris - Comments: Listed as

"Black" in *A List of Pensioners of the War of 1812*.

Minot, Samuel - Private - 31st U.S. Infantry - Company: Ethan Burnap - Age: 22 - Height: 5' 9" - Eyes: Black - Hair: Black - Complexion: Black - Born: MA, Suffolk County, Boston - Trade: Farmer - Enlistment date: 3 Mar 1814 - Enlistment place: VT, Thetford or Montpelier - Enlistment period: War - Enlisted by whom: John Hatch - Discharge date: 4 Jun 1815 - Comments: Possible Black, not verified; discharged at Surgeon's Certificate of Discharge.

Moasley, William - Private - 39th U.S. Infantry - Company: George Hallum - Discharge date: 9 Jan 1815 - Comments: Reported as a negro on the regimental muster rolls completed after the Battle of Horseshoe Bend, AL.

Moore, John - Recruit - 26th U.S. Infantry - Recruiting Detachment: William Bezeau - Age: 30 - Height: 5' - Eyes: Black - Hair: Black - Complexion: Black - Born: PA, Philadelphia - Trade: Blacksmith - Enlistment date: 10 Sep 1814 - Enlistment place: Philadelphia - Enlistment period: 5 Yrs - Enlisted by whom: William Bezeau - Discharge date: 20 May 1815 - Comments: Listed as "Col'd" in his service record; discharged at Philadelphia, PA.

Moore, John - Private - 31st U.S. Infantry - Age: 21 - Height: 5' 5" - Eyes: Black - Hair: Black - Complexion: Black - Born: NH, Rockingham County, Londonderry - Trade: Barber - Enlistment date: 18 Apr 1814 - Enlistment place: VT, Pomfred or Thetford - Enlistment period: War - Enlisted by whom: Amos Brown - Died: 8 Dec 1814 - Comments: Possible Black, not verified.

Moore, Samuel (Moores) - Private - U.S. Light Artillery - Company: George Morris - Age: 21 - Height: 5' 6" - Eyes: Black - Hair: Black - Complexion: Black - Born: MA, Berkshire County, Lanesborough - Enlistment date: 15 Jul 1814 - Enlistment place: MA, Boston - Enlistment period: War - Enlisted by whom: Samuel Washburn - Comments: Possible Black, not verified; absent 16 Feb 1815, left sick at Plattsburgh, NY, supposed to be at Sackets Harbor, NY.

Moore, Silas - Private - 4th U.S. Infantry - Company: Ebenezer Way - Age: 21 - Height: 5' - Eyes: Black - Hair: Black - Complexion: Black - Born: NH, Hillsborough County, Dumbarton - Trade: Laborer - Enlistment date: 4 Oct 1814 - Enlistment period: War - Enlisted by whom: Lieutenant Low - Discharge date: 1 Apr 1815 - Comments: Possible Black, not verified; discharged at Portsmouth, NH.

Morris, Samuel - Recruit - 26th U.S. Infantry - Recruiting Detachment: William Bezeau - Age: 39 - Height: 5' 5" - Eyes: Black - Hair: Curly - Complexion: Black - Born: PA, Lancaster County - Trade: Sailor - Enlistment date: 10 Oct 1814 - Enlistment place: Philadelphia - Enlistment period: 5 Yrs - Enlisted by whom: William Bezeau - Discharge date: 20 May 1815 - Comments: Listed as "Col'd" in his service record; discharged at Philadelphia, PA.

Morton, Isaac - Recruit - 26th U.S. Infantry - Recruiting Detachment: William Bezeau - Age: 34 - Height: 5' 9" - Eyes: Black - Hair: Curly - Complexion: Black - Born: Pennsylvania - Trade: Farmer - Enlistment date: 11 Feb 1815 - Enlistment place: Philadelphia - Enlistment period: War - Enlisted by whom: William Bezeau - Comments: Possible Black, not verified.

Munn, Edward - Recruit - 26th U.S. Infantry - Recruiting Detachment: William Bezeau - Age: 22 - Height: 5' 10" - Eyes: Black - Hair: Curly - Complexion: Black - Born: Pennsylvania - Trade: Farmer - Enlistment date: 23 Jan 1815 - Enlistment place: Philadelphia - Enlistment period: War - Enlisted by whom: William Bezeau - Comments: Listed on the Descriptive Roll of Colored Men, 1 Apr 1815; deserted on 27 Jan 1815.

Murrey, Joseph - Recruit - U.S. Light Artillery - Age: 20 - Height: 5' 6" - Born: Spain, Havana (Cuba) - Enlistment date: 9 Jul 1814 - Enlistment place: MA, Boston - Enlistment period: War - Enlisted by whom: Samuel Washburn - Comments: Possible Black, not verified.

Nahn, John - Recruit - 26th U.S. Infantry - Recruiting Detachment: William Bezeau - Age: 21 - Height: 5' 3 1/2" - Eyes: Black - Hair: Curly - Complexion: Black - Born: Africa - Trade: Laborer - Enlistment date: 21 Nov 1814 - Enlistment place: Philadelphia - Enlistment period: War - Enlisted by whom: William Bezeau - Comments: Listed as "Col'd" in his service record; deserted on 22 Nov 1814.

Nelson, Isaac - Musician - U.S. Light Artillery - Company: George Morris - Age: 21 - Height: 5' 5" - Eyes: Black - Hair: Black - Complexion: Black - Born: NY, Albany County - Trade: Farmer - Enlistment date: 8 Sep 1814 - Enlistment place: MA, Boston - Enlistment period: War - Enlisted by whom: Samuel Washburn - Comments: Possible Black, not verified; discharged on 31 Mar 1815.

Newburn, John - Recruit - 26th U.S. Infantry - Recruiting Detachment: William Bezeau - Age: 28 - Height: 5' 9" - Eyes: Black - Hair: Curly - Complexion: Black - Born: Pennsylvania - Trade: Laborer - Enlistment date: 17 Jan 1815 - Enlistment place: Philadelphia - Enlistment period: War - Enlisted by whom: William Bezeau - Comments: Listed on the Descriptive Roll of Colored Men, 1 Apr 1815.

Nichols, Samuel - Recruit - 26th U.S. Infantry - Recruiting Detachment: William Bezeau - Age: 28 - Height: 5' 7" - Eyes: Black - Hair: Curly - Complexion: Brown - Born: Virginia - Trade: Laborer - Enlistment date: 1 Jan 1815 - Enlistment period: War - Bounty: BLW 537-320-14 (Illinois) - Comments: Possible Black, not verified; discharged at Philadelphia, PA.

Nixon, Richard - Recruit - 26th U.S. Infantry - Recruiting Detachment: William Bezeau - Age: 25 - Height: 5' - Eyes: Black - Hair: Curly - Complexion: Chestnut - Born: New York - Trade: Laborer - Enlistment date: 2 Nov 1814 - Enlistment place: Philadelphia - Enlistment period: War - Enlisted by whom: William Bezeau - Discharge date: 20 Mar 1815 - Bounty: BLW 7289-160-12 (Illinois) - Comments: Listed as "Col'd" in his service record; discharged at Philadelphia, PA.

Noble, Jordan B. - Musician - 7th U.S. Infantry - Pension: SO-22074, SC-19527, IO-408304; served as a musician in the 7th U.S. Infantry (War of 1812), in Colonel P. F. Smith's Louisiana Volunteers (Florida War of 1836), in Colonel Watrous' Louisiana Volunteers (Mexican War), and as a captain of Company C, Louisiana Volunteers (Civil War) - Bounty: BLW 30300-160-55 (Mexican War Service) - Comments: Service record has not been found; received pension for Mexican War service.

Nopley, George - Recruit - 26th U.S. Infantry - Recruiting Detachment: William Bezeau - Age: 34 - Height: 6' - Eyes: Black - Hair: Curly - Complexion: Brown - Born: Pennsylvania - Trade: Laborer - Enlistment date: 20 Jan 1815 - Enlistment place: Philadelphia - Enlistment period: War - Enlisted by whom: William Bezeau - Comments: Listed on the Descriptive Roll of Colored Men, 1 Apr 1815.

Norlin, Luke - Recruit - 26th U.S. Infantry - Recruiting Detachment: William Bezeau - Age: 35 - Height: 5' 7" - Eyes: Black - Hair: Curly - Complexion: Black - Born: Pennsylvania - Trade: Laborer - Enlistment date: 16 Jan 1815 - Enlistment place: Philadelphia - Enlistment period: War - Enlisted by whom: William Bezeau - Comments: Listed on the Descriptive Roll of Colored Men, 1 Apr 1815.

Obman, Edward - Recruit - 26th U.S. Infantry - Recruiting Detachment: William Bezeau - Age: 30 - Height: 5' 7" - Eyes: Black - Hair: Curly - Complexion: Black - Born: Pennsylvania - Trade: Laborer - Enlistment date: 22 Jan 1815 - Enlistment place: Philadelphia - Enlistment period: War - Enlisted by whom: William Bezeau - Comments: Listed on the Descriptive Roll of Colored Men, 1 Apr 1815; deserted on 27 Jan 1815.

Olden, Seth - Recruit - 26th U.S. Infantry - Recruiting Detachment: William Bezeau - Age: 19 - Height: 5' 5" - Eyes: Black - Hair: Black - Complexion: Black - Born: PA, Philadelphia - Trade: Laborer - Enlistment date: 3 Sep 1814 - Enlistment place: Philadelphia - Enlistment period: 5 Yrs - Enlisted by whom: William Bezeau - Discharge date: 16 May 1815 - Comments: Listed as "Col'd" in his service record; discharged at Philadelphia, PA.

Oston, Charles - Recruit - 26th U.S. Infantry - Recruiting Detachment: William Bezeau - Age: 24 - Height: 5' 6" - Eyes: Black - Hair: Curly - Complexion: Black - Born: Pennsylvania - Trade: Laborer - Enlistment date: 4 Jan 1815 - Enlistment place: Philadelphia - Enlistment period: War - Enlisted by whom: William Bezeau - Comments: Listed on the Descriptive Roll of Colored Men, 1 Apr 1815.

Ostrander, Harry - Private - 29th U.S. Infantry - Company: Mathew Danvers - Age: 23 - Height: 5' 7 1/2" - Eyes: Black - Hair: Black - Complexion: Yellow - Born: NY, Albany County, Waterfleet - Trade: Laborer - Enlistment date: 8 Dec 1814 - Enlistment place: NY, Troy - Enlistment period: War - Enlisted by whom: Thomas Vail - Discharge date: 29 Jun 1815 - Comments: Possible Black, not verified; discharged at Champlain.

Page, Sampson - Private - U.S. Light Artillery - Company: John McIntosh - Age: 29 - Height: 5' 3 1/2" - Eyes: Black - Hair: Black - Complexion: Black - Born: NH, Hillsborough County, Dumbarton - Trade: Farmer - Enlistment date: 20 Jul 1814 - Enlistment place: MA, Boston - Enlistment period: War - Enlisted by whom: Samuel Washburn - Discharge date: 31 May 1815 - Comments: Possible Black, not verified; discharged at Plattsburgh, NY.

Painter, William - Recruit - 26th U.S. Infantry - Recruiting Detachment: William Bezeau - Age: 24 - Height: 5' 4" - Eyes: Black - Hair: Curly - Complexion: Yellow - Born: Pennsylvania - Trade: Seaman - Enlistment date: 3 Jan 1815 - Enlistment place: Philadelphia - Enlistment period: War - Enlisted by whom: William Bezeau - Comments: Listed on the Descriptive Roll of Colored Men, 1 Apr 1815.

Palmer, Jacob - Private - 31st U.S. Infantry - Company: Rufus Stewart - Age: 24 - Height: 5' 11" - Eyes: Black - Hair: Black - Complexion: Black - Born: CT, New London County, Stonington - Trade: Farmer - Enlistment date: 22 Feb 1814 - Enlistment place: NY, Plattsburgh - Enlistment period: War - Enlisted by whom: Ethan Burnap - Comments: Possible Black, not verified; deserted from Plattsburgh, NY, on 9 Jul 1814.

Palmer, John - Recruit - 26th U.S. Infantry - Recruiting Detachment: William Bezeau - Age: 29 - Height: 5' 6" - Eyes: Black - Hair: Curly - Complexion: Brown - Born: New Jersey - Trade: Laborer - Enlistment date: 20 Nov 1814 - Enlistment place: Philadelphia - Enlistment period: War - Enlisted by whom: William Bezeau - Discharge date: 20 Mar 1815 - Bounty: BLW 7204-160-12 (Illinois) - Comments: Listed as "Colored" in his service record; discharged at Philadelphia, PA.

Parker, John - Recruit - 26th U.S. Infantry - Recruiting Detachment: William Bezeau - Age: 25 - Height: 5' 9" - Born: Pennsylvania - Enlistment date: 21 Sep 1814 - Enlistment place: Philadelphia -

Enlisted by whom: William Bezeau - Comments: Listed as "Colored" in his service record; deserted on 28 Sep 1814.

Paskal, John - Recruit - 26th U.S. Infantry - Recruiting Detachment: William Bezeau - Age: 28 - Height: 5' 5" - Eyes: Black - Hair: Curly - Complexion: Black - Born: Delaware - Trade: Sailor - Enlistment date: 2 Nov 1814 - Enlistment place: Philadelphia - Enlistment period: War - Enlisted by whom: William Bezeau - Discharge date: 20 Mar 1815 - Bounty: BLW 10290-160-12 (Illinois) - Comments: Listed as "Col'd" in his service record; discharged at Philadelphia, PA.

Paul, John - Private - U.S. Light Artillery - Company: John McIntosh - Age: 23 - Height: 5' 6" - Eyes: Black - Hair: Black - Complexion: Black - Born: LA, New Orleans - Trade: Mariner - Enlistment date: 15 Jul 1814 - Enlistment place: MA, Boston - Enlistment period: War - Enlisted by whom: Samuel Washburn - Comments: Possible Black, not verified; absent 16 Feb 1815, left sick at Plattsburgh, NY.

Perry, James - Private - U.S. Light Artillery - Company: John Bell - Age: 18 - Height: 5' 6" - Eyes: Black - Hair: Black - Complexion: Black - Born: Boston, Baltimore or Queen Ann County - Trade: Mariner - Enlistment date: 3 Nov 1814 - Enlistment place: MA, Concord or Boston - Enlistment period: 5 Yrs - Enlisted by whom: Richard Field - Discharge date: 31 Mar 1815 - Comments: Discharged on account of being a Negro and not a fit companion for the American soldier.

Peters, Jacob - Recruit - 26th U.S. Infantry - Recruiting Detachment: William Bezeau - Age: 36 - Height: 5' 5" - Eyes: Dark - Hair: Dark - Complexion: Dark - Born: Pennsylvania - Trade: Laborer - Enlistment date: 22 Nov 1814 - Enlistment place: Philadelphia - Enlisted by whom: William Bezeau - Discharge date: 20 Mar 1815 - Bounty: BLW 18555-160-12 (Illinois) - Comments: Listed as "Colored" in his service record; discharged at Philadelphia, PA.

Peters, John - Recruit - 26th U.S. Infantry - Recruiting Detachment: William Bezeau - Age: 30 - Height: 5' 7" - Eyes: Black - Hair: Black - Complexion: Black - Born: Saint Domingo (Haiti) - Trade: Cook - Enlistment date: 22 Nov 1814 - Enlistment place: Philadelphia - Enlistment period: 5 Yrs - Enlisted by whom: William Bezeau - Discharge date: 20 Mar 1815 - Comments: Listed on the Descriptive Roll of Negroes, 25 May 1815 and as "Colored" in his service record; discharged at Philadelphia, PA.

Peters, Scipio - Recruit - 26th U.S. Infantry - Recruiting Detachment: William Bezeau - Age: 40 - Height: 5' 7" - Born: PA, Philadelphia - Enlistment date: 24 Nov 1814 - Enlistment place: Philadelphia - Enlistment period: 5 Yrs - Enlisted by whom: William Bezeau - Comments: Listed as "Colored" in his service record; deserted on 10 Dec 1814.

Phillips, George - Private - U.S. Light Artillery - Company: George Morris - Age: 26 - Height: 5' 6 1/2" - Eyes: Black - Hair: Black - Complexion: Black - Born: MA, Boston or Lexington - Trade: Farmer - Enlistment date: 3 Oct 1814 - Enlistment place: MA, Boston - Enlistment period: War - Enlisted by whom: William Campbell - Discharge date: 31 Mar 1815 - Comments: Possible Black, not verified.

Phillips, George - Recruit - 26th U.S. Infantry - Recruiting Detachment: William Bezeau - Age: 30 - Height: 5' 7" - Eyes: Black - Hair: Black - Complexion: Black - Born: Pennsylvania - Trade: Laborer - Enlistment date: 31 Oct 1814 - Enlistment place: Philadelphia - Enlistment period: 5 Yrs - Enlisted by whom: William Bezeau - Discharge date: 1 May 1815 - Comments: Listed as a "Colored man" in his service record; discharged at Philadelphia, PA.

Phillips, Isaac - Recruit - 26th U.S. Infantry - Recruiting Detachment: William Bezeau - Age: 22 - Height: 5' 5 1/2" - Eyes: Hazel - Hair: Curly - Complexion: Yellow - Born: Pennsylvania - Trade: Tobacconist - Enlistment date: 5 Nov 1814 - Enlistment place: Philadelphia - Enlistment period: War - Enlisted by whom: William Bezeau - Discharge date: 23 Mar 1815 - Bounty: BLW 14820-160-12 (Illinois) - Comments: Listed as "Colored" in his service record; discharged at Philadelphia, PA.

Pierce, James - Recruit - 26th U.S. Infantry - Recruiting Detachment: William Bezeau - Age: 20 - Height: 5' 7 1/2" - Born: Delaware - Enlistment date: 23 Nov 1814 - Enlistment place: Philadelphia - Enlistment period: 5 Yrs - Enlisted by whom: William Bezeau - Comments: Listed as "Colored" in his service record; given up to the Navy, Jan 1815, a deserter.

Pierce, Lewis - Recruit - 26th U.S. Infantry - Recruiting Detachment: William Bezeau - Age: 20 - Height: 5' 2" - Eyes: Black - Hair: Curly - Complexion: Black - Born: Pennsylvania - Trade: Laborer - Enlistment date: 8 Feb 1815 - Enlistment place: Philadelphia - Enlistment period: War - Enlisted by whom: William Bezeau - Comments: Listed on the Descriptive Roll of Colored Men, 1 Apr 1815.

Pitts, Jesse - Private - 8th U.S. Infantry - Company: Thomas Farrar - Age: 23 - Height: 5' 9 1/4" - Eyes: Black - Hair: Black - Complexion: Brown - Born: SC, Barnwell County - Trade: Farmer - Enlistment date: 11 Aug 1814 - Enlistment place: GA, Savannah, Camp Flourney - Enlistment period: 5 Yrs - Enlisted by whom: William Bee - Comments: Possible Black, not verified.

Plaine, Samuel (Plain) - Recruit (Corporal) - 26th U.S. Infantry - Recruiting Detachment: William Bezeau - Age: 22 - Height: 5' 9 1/2" - Eyes: Brown - Hair: Brown - Complexion: Brown - Born: Pennsylvania - Trade: Cooper - Enlistment date: 8 Sep 1814 - Enlistment place: Philadelphia - Enlistment period: War - Enlisted by whom: William Bezeau - Discharge date: 23 Mar 1815 - Bounty: BLW 10297-160-42 (Illinois) - Comments: Possible Black, not verified; discharged at Philadelphia, PA.

Pomroy, William - Recruit - 26th U.S. Infantry - Recruiting Detachment: William Bezeau - Age: 32 - Height: 5' 3" - Eyes: Black - Hair: Curly - Complexion: Black - Born: Pennsylvania - Trade: Farmer - Enlistment date: 7 Jan 1814 - Enlistment place: Philadelphia - Enlistment period: War - Enlisted by whom: William Bezeau - Comments: Listed on the Descriptive Roll of Colored Men, 1 Apr 1815.

Portland, Thomas - Recruit - 26th U.S. Infantry - Recruiting Detachment: William Bezeau - Age: 17 - Height: 4' 8" - Eyes: Brown - Hair: Curly - Complexion: Black - Born: New Jersey - Trade: Waiter - Enlistment date: 4 Dec 1814 - Enlistment period: War - Enlisted by whom: William Bezeau - Discharge date: 20 Mar 1815 - Bounty: BLW 5889-160-12 (Illinois) - Comments: Listed as "Colored" in his service record, discharged at Philadelphia, PA.

Prentis, Leonard (Prentice) - Private - U.S. Light Artillery - Company: George Morris - Age: 21 - Height: 5' 5 3/4" - Eyes: Black - Hair: Black or Dark - Complexion: Black - Born: NH, Rockingham County, Londonderry - Trade: Mariner - Enlistment date: 18 Jul 1814 - Enlistment place: MA, Boston - Enlistment period: War - Enlisted by whom: Samuel Washburn - Discharge date: 31 Mar 1815 - Comments: Possible Black, not verified.

Pullard, Jacob - Recruit - 26th U.S. Infantry - Recruiting Detachment: William Bezeau - Age: 22 - Height: 5' 6" - Eyes: Black - Hair: Curly - Complexion: Black - Born: Pennsylvania - Trade: Laborer - Enlistment date: 15 Jan 1815 - Enlistment place: Philadelphia - Enlistment period: War - Enlisted by whom: William Bezeau - Comments: Listed on the Descriptive Roll of Colored Men, 1

Apr 1815.

Pursue, Aaron - Recruit - 26th U.S. Infantry - Recruiting Detachment: William Bezeau - Age: 32 - Height: 5' - Eyes: Hazel - Hair: Curly - Complexion: Yellow - Born: New Jersey - Trade: Laborer - Enlistment date: 14 Oct 1814 - Enlistment place: Philadelphia - Enlistment period: War - Enlisted by whom: William Bezeau - Discharge date: 23 Mar 1815 - Bounty: BLW 17203-160-42 (Illinois) - Comments: Listed as "Col'd" in his service record; discharged at Philadelphia, PA.

Ramsey, Joseph - Recruit - 26th U.S. Infantry - Recruiting Detachment: William Bezeau - Age: 17 - Height: 5' 4" - Eyes: Curly - Hair: Curly - Complexion: Black - Born: Pennsylvania - Trade: Laborer - Enlistment date: 2 Feb 1815 - Enlistment place: Philadelphia - Enlistment period: War - Enlisted by whom: William Bezeau - Comments: Listed on the Descriptive Roll of Colored Men, 1 Apr 1815.

Rensellar, John (Ransalier) - Private - 11th U.S. Infantry - Company: Benjamin Smead - Age: 26 - Height: 5' 6" - Eyes: Black - Hair: Black - Complexion: Black - Trade: Farmer - Enlistment date: 4 Mar 1813 - Enlistment period: 5 Yrs - Comments: Listed as a "Blackman" in his service record; American prisoner of war from Halifax, NS.

Richardson, John - Recruit - 26th U.S. Infantry - Recruiting Detachment: William Bezeau - Age: 27 - Height: 5' 10 1/2" - Eyes: Black - Hair: Black - Complexion: Yellow - Born: Virginia - Trade: Laborer - Enlistment date: 1 Nov 1814 - Enlistment place: Philadelphia - Enlistment period: War - Enlisted by whom: William Bezeau - Discharge date: 22 Mar 1815 - Bounty: BLW 7291-120-12 (Illinois) - Comments: Listed as "Col'd" in his service record; discharged at Philadelphia, PA.

Ridgeway, Samuel - Laborer - Corps of Artificers - Company: Alexander Paris - Comments: Listed as "Black" in *A List of Pensioners of the War of 1812*.

Roberts, Thomas - Recruit - 26th U.S. Infantry - Recruiting Detachment: William Bezeau - Age: 19 - Height: 5' 3" - Eyes: Black - Hair: Curly - Complexion: Black - Born: Pennsylvania - Trade: Laborer - Enlistment date: 17 Jan 1815 - Enlistment place: Philadelphia - Enlistment period: War - Enlisted by whom: William Bezeau - Comments: Listed on the Descriptive Roll of Colored Men, 1 Apr 1815.

Robinson, George (Robertson) - Private - U.S. Light Artillery - Company: George Morris - Age: 28 - Height: 5' 5 3/4" - Eyes: Black - Hair: Black - Complexion: Black - Born: VA, Accomack County - Trade: Mariner - Enlistment date: 6 Oct 1814 - Enlistment place: MA, Boston - Enlistment period: War - Enlisted by whom: William Campbell - Comments: Possible Black, not verified.

Robinson, Simon (Robertson) - Private - U.S. Light Artillery - Company: George Morris - Age: 32 - Height: 6' - Eyes: Black - Hair: Black - Complexion: Black or Dark - Born: New Jersey - Trade: Farmer - Enlistment date: 1 Aug 1814 - Enlistment place: MA, Boston - Enlistment period: War - Enlisted by whom: Samuel Washburn - Discharge date: 31 Mar 1815 - Comments: Possible Black, not verified.

Rutter, Stephen - Private - U.S. Light Artillery - Company: John McIntosh - Age: 29 - Height: 5' 5" - Eyes: Black - Hair: Black - Complexion: Black - Born: PA, Philadelphia - Trade: Mariner - Enlistment date: 11 Jul 1814 - Enlistment place: MA, Boston - Enlistment period: War - Enlisted by whom: Samuel Washburn - Discharge date: 31 May 1815 - Comments: Possible Black, not verified; discharged at Plattsburgh, NY.

Sanderson, James - Recruit - 26th U.S. Infantry - Recruiting Detachment: William Bezeau - Age: 22 - Height: 5' 11" - Eyes: Black - Hair: Curly - Complexion: Chestnut - Born: New Jersey - Trade: Laborer - Enlistment date: 3 Nov 1814 - Enlistment place: Philadelphia - Enlistment period: War - Enlisted by whom: William Bezeau - Comments: Listed as "Col'd" in his service record.

Sawyer, Nathaniel - Laborer - Corps of Artificers - Company: Alexander Paris - Comments: Listed as "Black" in *A List of Pensioners of the War of 1812*.

Scott, Francis - Private - 39th U.S. Infantry - Company: George Hallum - Enlistment date: 23 Jan 1814 - Discharge date: 22 Jan 1815 - Comments: Reported as a negro on the regimental muster rolls completed after the Battle of Horseshoe Bend.

Scott, Wade - Private - 24th U.S. Infantry - Company: James Campbell - Age: 35 - Height: 5' 10" - Eyes: Yellow - Hair: Curly - Complexion: Yellow - Born: Virginia - Trade: Farmer - Enlistment date: 20 Jan 1813 - Enlistment place: Wilson County, TN - Enlistment period: 5 Yrs - Enlisted by whom: Lt Hays - Comments: A Negro.

Seamore, John - Recruit - U.S. Light Artillery - Age: 21 - Height: 5' 2" - Born: Africa - Enlistment date: 18 Jul 1814 - Enlistment place: MA, Boston - Enlisted by whom: Samuel Washburn - Comments: Possible Black, not verified.

Shanklin, Myers (or Mayer) - Recruit - 26th U.S. Infantry - Recruiting Detachment: William Bezeau - Age: 18 - Height: 5' 6" - Eyes: Black - Hair: Curly - Complexion: Black - Born: Pennsylvania - Trade: Rope maker - Enlistment date: 28 Dec 1814 - Enlistment place: Philadelphia - Enlistment period: 5 Yrs - Enlisted by whom: William Bezeau - Discharge date: 22 May 1815 - Comments: Listed on the Descriptive Roll of Negroes, 25 May 1815 and as "Col'd" in his service record; discharged at Philadelphia, PA.

Sherbourne, William (Shurbarn) - Private - U.S. Light Artillery - Company: John McIntosh - Age: 29 - Height: 5' 6 1/2" - Eyes: Black - Hair: Black - Complexion: Black - Born: NH, Rockingham County, Londonderry - Trade: Mariner - Enlistment date: 17 Jul 1814 - Enlistment place: MA, Boston - Enlistment period: War - Enlisted by whom: Samuel Washburn - Discharge date: 25 May 1815 - Bounty: BLW 13924-160-12 (Illinois) - Comments: Possible Black, not verified; discharged at Sackets Harbor, NY.

Simmons, John - Private - U.S. Light Artillery - Company: George Morris - Age: 23 - Height: 5' 6 1/2" - Eyes: Black - Hair: Black - Complexion: Black - Born: Bermuda - Trade: Mariner - Enlistment date: 8 Oct 1814 - Enlistment place: MA, Boston - Enlistment period: War - Enlisted by whom: William Campbell - Discharge date: 31 Mar 1815 - Comments: Possible Black, not verified.

Simpson, Peter - Private - 23rd U.S. Infantry - Company: Justus Ingersoll - Age: 24 - Height: 5' 6" - Eyes: Mulatto - Hair: Mulatto - Complexion: Mulatto - Born: CT, Windham County, Brookfield - Trade: Shoemaker - Enlistment date: 7 Jan 1814 - Enlistment place: Batstown - Enlistment period: 5 Yrs - Enlisted by whom: Richard Smyth - Discharge date: 8 Jan 1815 - Bounty: BLW 9594-160-42 (Illinois) - Comments: Listed as a "Mulatto" in his service record; wounded at Fort Erie, UC; discharged at Sackets Harbor, NY.

Smith, John - Recruit - 26th U.S. Infantry - Recruiting Detachment: William Bezeau - Age: 24 - Height: 5' 6" - Eyes: Black - Hair: Curly - Complexion: Black - Born: New Jersey - Trade: Laborer - Enlistment date: 1 Jan 1815 - Enlistment place: Philadelphia - Enlistment period: 5 Yrs - Enlisted by whom: William Bezeau - Discharge date: 1 May 1815 - Comments: Listed as a "Blackman" in his service record; discharged at Philadelphia, PA.

Smith, Solomon - Recruit - 26th U.S. Infantry - Recruiting Detachment: William Bezeau - Age: 33 - Height: 5' 9" - Eyes: Black - Hair: Black - Complexion: Black - Born: Delaware - Trade: Laborer - Enlistment date: 30 Nov 1814 - Enlistment place: Philadelphia - Enlistment period: 5 Yrs - Enlisted by whom: William Bezeau - Discharge date: 20 May 1815 - Comments: Listed as a "Colored man" in his service record.

Smith, Thomas P. - Private - 23rd U.S. Infantry - Company: Azariah Odell - Other regiment: 2nd U.S. Infantry - Age: 38 - Height: 5' 8" - Eyes: Black - Hair: Black - Complexion: Black - Born: PA, Allegheny County, Pittsburgh - Trade: Sawyer - Enlistment date: 27 Jun 1814 - Enlistment place: NY, Albany - Enlistment period: War - Enlisted by whom: William Clarke - Discharge date: 5 Jun 1815 - Comments: Possible Black, not verified.

Smith, William - Recruit - 26th U.S. Infantry - Recruiting Detachment: William Bezeau - Age: 22 - Height: 5' 5" - Eyes: Black - Hair: Curly or Black - Complexion: Yellow - Born: Pennsylvania - Trade: Farmer - Enlistment date: 11 Nov 1814 - Enlistment place: Philadelphia - Enlistment period: 5 Yrs - Enlisted by whom: William Bezeau - Discharge date: 17 May 1815 - Comments: Listed on the Descriptive Roll of Negroes, 25 May 1815 and as "Col'd" in his service record; discharged at Philadelphia, PA.

Spanason, John - Recruit - 26th U.S. Infantry - Recruiting Detachment: William Bezeau - Age: 22 - Height: 5' 10" - Eyes: Black - Hair: Curly - Complexion: Brown - Born: PA, Philadelphia - Trade: Laborer - Enlistment date: 21 Nov 1814 - Enlistment place: Philadelphia - Enlistment period: War - Enlisted by whom: William Bezeau - Comments: Listed as "Col'd" in his service record.

Spencer, Benjamin - Recruit - 26th U.S. Infantry - Recruiting Detachment: William Bezeau - Age: 28 - Height: 5' 5" - Eyes: Black - Hair: Curly - Complexion: Black - Born: Delaware - Trade: Laborer - Enlistment date: 3 Jan 1815 - Enlistment place: Philadelphia - Enlistment period: War - Enlisted by whom: William Bezeau - Bounty: BLW 812-320-14 Cancelled; BLW 1038-320-14 - Comments: Listed on the Descriptive Roll of Colored Men, 1 Apr 1815; land bounty to Isaac Spencer, cousin & only heir at law of Benjamin Spencer, deceased; died before 6 Jul 1819; first bounty cancelled by an order from the War Department on 19 May 1820.

Stanley, Samuel - Private - 11th U.S. Infantry - Company: Horace Hale - Age: 15 - Height: 5' 7 1/2" - Eyes: Black - Hair: Black - Complexion: Black - Born: MA, Oxford County, Shapley - Trade: Farmer - Enlistment date: 6 Feb 1813 - Enlistment place: Pottersfield - Enlistment period: 18 Months - Enlisted by whom: Jonathan Stark - Discharge date: 10 Aug 1814 - Comments: Possible Black, not verified; discharged at Fort Erie, UC.

Stratten, Isaac (Statten) - Recruit - 26th U.S. Infantry - Recruiting Detachment: William Bezeau - Age: 26 - Height: 5' 5" - Eyes: Black - Hair: Black - Complexion: Black - Born: New Jersey - Trade: Shoemaker - Enlistment date: 21 Sep 1814 - Enlistment place: Philadelphia - Enlistment period: 5 Yrs - Enlisted by whom: William Bezeau - Discharge date: 25 May 1815 - Comments: Listed on the Descriptive Roll of Negroes, 25 May 1815 and as "Col'd" in his service record.

Tallsbury, Caesar (Tallsberry) - Private - U.S. Light Artillery - Company: George Morris - Age: 19 - Height: 5' 3" - Eyes: Black - Hair: Black - Complexion: Black - Born: RI, Warwick County, Coventry - Trade: Laborer - Enlistment date: 29 Jul 1814 - Enlistment place: RI, Providence County, Dedham - Enlistment period: War - Enlisted by whom: William Campbell - Discharge date: 31 Mar 1815 - Comments: Listed as "Blk" in his service record.

Taylor, Isaac - Recruit - 26th U.S. Infantry - Recruiting Detachment: William Bezeau - Age: 24 - Height: 5' 6" - Eyes: Black - Hair: Curly - Complexion: Brown - Born: Delaware - Trade: Farmer - Enlistment date: 27 Jan 1815 - Enlistment place: Philadelphia - Enlistment period: War - Enlisted by whom: William Bezeau - Comments: Listed on the Descriptive Roll of Colored Men, 1 Apr 1815.

Taylor, Julius C. - Recruit (Sergeant) - 26th U.S. Infantry - Recruiting Detachment: William Bezeau - Age: 32 - Height: 6' - Eyes: Grey - Hair: Brown - Complexion: Fair or Light - Born: NY, Columbia County - Trade: Farmer - Enlistment date: 12 Feb 1815 - Enlistment period: War - Enlisted by whom: William Bezeau - Discharge date: 30 May 1815 - Bounty: BLW 212-320-14 - Comments: Listed as "Col'd" in his service record; discharged at Philadelphia, PA.

Thomas, James - Recruit - 26th U.S. Infantry - Recruiting Detachment: William Bezeau - Age: 17 - Height: 5' 4" - Eyes: Black - Hair: Curly - Complexion: Dark - Born: Pennsylvania - Trade: Laborer - Enlistment date: 16 Nov 1814 - Enlistment place: Philadelphia - Enlistment period: War - Enlisted by whom: William Bezeau - Discharge date: 20 Mar 1815 - Bounty: BLW 10298-160-12 (Illinois) - Comments: Listed as "Col'd" in his service record; discharged at Philadelphia, PA.

Thomas, Robert - Laborer - Corps of Artificers - Company: Alexander Paris - Comments: Listed as "Black" in *A List of Pensioners of the War of 1812*.

Thompson, Francis - Private - 11th U.S. Infantry - Company: William Foster - Age: 21 - Height: 5' 8" - Eyes: Black - Hair: Black - Complexion: Black - Born: NY, Reussaleau County, Bethlehem - Trade: Farmer - Enlistment date: 25 Mar 1813 - Enlistment place: VT, Bennington - Enlistment period: 5 Yrs - Enlisted by whom: Charles Follett - Comments: Possible Black, not verified; prisoner of War, disabled from wounds.

Thomson, Abram - Private - 31st U.S. Infantry - Age: 27 - Height: 5' 6" - Eyes: Black - Hair: Black - Complexion: Black - Born: NY, New York - Trade: Farmer - Enlistment date: 2 Nov 1814 - Enlistment place: VT, Guilford or Brattleboro - Enlistment period: War - Enlisted by whom: Jeremiah Greenleaf - Discharge date: 31 Mar 1815 - Bounty: BLW 4995-160-12 (Illinois) - Comments: Possible Black, not verified; discharged at Woodstock, VT; listed as Abraham Thompson on his land bounty warrant.

Toby, James - Recruit - 26th U.S. Infantry - Recruiting Detachment: William Bezeau - Age: 31 - Height: 5' 10 1/2" - Eyes: Black - Hair: Curly - Complexion: Brown - Born: Delaware - Trade: Farmer - Enlistment date: 19 Jan 1815 - Enlistment place: Philadelphia - Enlistment period: War - Enlisted by whom: William Bezeau - Comments: Probably Black, not verified; listed on the Descriptive Roll of 1 Apr 1815; deserted on 24 Jan 1815.

Townsend, James - Private - U.S. Light Artillery - Company: John McIntosh - Age: 20 - Height: 5' 7" - Eyes: Black - Hair: Black - Complexion: Black - Born: NY, Newson - Trade: Laborer - Enlistment date: 18 Jul 1814 - Enlistment place: MA, Boston - Enlistment period: War - Enlisted by whom: Samuel Washburn - Discharge date: 31 May 1815 - Comments: Possible Black, not verified; discharged at Plattsburgh, NY.

Townsend, Jonathan - Laborer - Corps of Artificers - Company: Alexander Paris - Comments: Listed as "Black" in *A List of Pensioners of the War of 1812*.

Tredwell, Cyrus – Musician – 27th U.S. Infantry – Company: Allen Reynolds – Age: 21 – Height: 5' 9"

– Eyes: Hazel – Hair: Black – Complexion: Yellow or Black – Born: NY, Long Island – Trade: Musician – Enlistment date: 16 Sep 1814 – Enlistment period: War – Enlistment Place: New York – Enlisted by whom: Allen Reynolds – Discharge date: 20 Jun 1815.

Trusty, Campbell - Recruit - 26th U.S. Infantry - Recruiting Detachment: William Bezeau - Age: 23 - Height: 6' 1" - Eyes: Black - Hair: Curly - Complexion: Black - Born: Delaware - Trade: Laborer - Enlistment date: 3 Nov 1814 - Enlistment place: Philadelphia - Enlistment period: War - Enlisted by whom: William Bezeau - Discharge date: 23 Mar 1815 - Bounty: BLW 5296-160-12 (Illinois) - Comments: Listed as "Colored" in his service record; discharged at Philadelphia, PA.

Turner, Caesar - Recruit - 26th U.S. Infantry - Recruiting Detachment: William Bezeau - Age: 40 - Height: 5' 8" - Born: Pennsylvania - Enlistment date: 29 Nov 1814 - Enlistment place: Philadelphia - Enlisted by whom: William Bezeau - Comments: Listed as "Col'd" in his service record.

Valloon, Abraham - Private - 11th U.S. Infantry - Company: William Foster - Other regiment: 6th U.S. Infantry - Age: 24 - Height: 5' 10" - Eyes: Black - Hair: Black - Complexion: Black - Born: VT, Essex County, Lunenburg - Trade: Farmer - Enlistment date: 13 May 1814 - Enlistment place: VT, Bennington - Enlistment period: 5 Yrs - Enlisted by whom: John Walker - Comments: Possible Black, not verified; discharged before 30 Jun 1816.

Van Kamp, Joseph - Recruit - 26th U.S. Infantry - Recruiting Detachment: William Bezeau - Age: 21 - Height: 5' 85' 8" - Eyes: Black - Hair: Black - Complexion: Brown - Born: Pennsylvania - Trade: Laborer - Enlistment date: 25 Oct 1814 - Enlistment place: Philadelphia - Enlistment period: War - Enlisted by whom: William Bezeau - Discharge date: 20 Mar 1815 - Bounty: BLW 7209-160-12 (Illinois) - Comments: Listed as "Col'd" in his service record; discharged at Philadelphia, PA.

Vineburner, Andrew (Venbunar) - Recruit - 26th U.S. Infantry - Recruiting Detachment: William Bezeau - Age: 34 - Height: 5' 8 1/2" - Eyes: Brown - Hair: Curly - Complexion: Yellow - Born: New York - Trade: Laborer - Enlistment date: 23 Nov 1814 - Enlistment place: Philadelphia - Enlistment period: War - Enlisted by whom: William Bezeau - Discharge date: 20 Mar 1815 - Bounty: BLW 51294-160-12 (Illinois) - Comments: Listed as "Col'd" in his service record; discharged at Philadelphia, PA.

Wallace, Warner (Walace) - Recruit - 26th U.S. Infantry - Recruiting Detachment: William Bezeau - Age: 19 - Height: 5' 8 1/2" – Eyes: Black - Hair: Curly - Complexion: Black - Born: New Jersey - Trade: Farmer - Enlistment date: 26 Dec 1814 - Enlistment place: Philadelphia - Enlistment period: War - Enlisted by whom: William Bezeau - Died: 14 Mar 1815 - Comments: Listed as "Col'd" in his service record; died in general hospital.

Wanton, Thomas - Private - U.S. Light Artillery - Company: George Morris - Age: 18 - Height: 5' 6" - Eyes: Black - Hair: Black - Complexion: Black - Born: RI, Washington County, Narragansett - Trade: Farmer - Enlistment date: 15 Aug 1814 - Enlistment place: RI, Providence - Enlistment period: War - Enlisted by whom: William Campbell - Discharge date: 31 Mar 1815 - Comments: Possible Black, not verified.

Weldrum, Andrew (Weldron) - Recruit - 26th U.S. Infantry - Recruiting Detachment: William Bezeau - Age: 31 - Height: 5' 6" - Eyes: Brown - Hair: Brown - Complexion: Yellow - Born: PA, Philadelphia - Trade: Farmer - Enlistment date: 3 Sep 1814 - Enlistment place: Philadelphia - Enlistment period: 5 Yrs - Enlisted by whom: William Bezeau - Discharge date: 13 May 1815 - Comments: Listed as a Negro in his service record; discharged at Philadelphia, PA.

Welton, Charles - Recruit - 26th U.S. Infantry - Recruiting Detachment: William Bezeau - Age: 21 - Height: 5' 4" - Eyes: Black - Hair: Curly - Complexion: Brown - Born: Pennsylvania - Trade: Farmer - Enlistment date: 9 Jan 1815 - Enlistment place: Philadelphia - Enlistment period: War - Enlisted by whom: William Bezeau - Comments: Probably Black, not verified; listed on the Descriptive Roll of 1 Apr 1815.

Welton, George - Recruit - 26th U.S. Infantry - Recruiting Detachment: William Bezeau - Age: 27 - Height: 5' 11" - Eyes: Black - Hair: Curly - Complexion: Black - Born: Delaware - Trade: Farmer - Enlistment date: 4 Feb 1815 - Enlistment place: Philadelphia - Enlistment period: War - Enlisted by whom: William Bezeau - Comments: Listed on the Descriptive Roll of Colored Men, 1 Apr 1815.

Wendell, Joseph (Wendall) - Private - U.S. Light Artillery - Company: John McIntosh - Age: 20 - Height: 5' 5" - Eyes: Black - Hair: Black - Complexion: Black - Born: MA, Suffolk County, Boston - Trade: Rope maker - Enlistment date: 15 Jul 1814 - Enlistment place: MA, Boston - Enlistment period: War - Enlisted by whom: Samuel Washburn - Discharge date: 31 May 1815 - Bounty: BLW 11680-160-42 (Illinois) - Comments: Possible Black, not verified; discharged at Plattsburgh, NY.

Wharton, Murray - Recruit - 26th U.S. Infantry - Recruiting Detachment: William Bezeau - Age: 22 - Height: 5' 8" - Eyes: Black - Hair: Curly - Complexion: Black - Born: Africa - Trade: Tanner - Enlistment date: 4 Oct 1814 - Enlistment place: Philadelphia - Enlistment period: War - Enlisted by whom: William Bezeau - Comments: Listed as "Col'd" in his service record; deserted on 30 Mar 1815; deserted from Lazaretto, PA.

Wheelock, Ezra - Laborer - Corps of Artificers - Company: Alexander Paris - Comments: Listed as "Black" in *A List of Pensioners of the War of 1812*.

White, Charles - Recruit - 26th U.S. Infantry - Recruiting Detachment: William Bezeau - Age: 21 - Height: 5' 5" - Eyes: Black - Hair: Curly - Complexion: Black - Born: Delaware - Trade: Laborer - Enlistment date: 4 Feb 1815 - Enlistment place: Philadelphia - Enlistment period: War - Enlisted by whom: William Bezeau - Comments: Listed on the Descriptive Roll of Colored Men, 1 Apr 1815.

White, Francis - Recruit - 26th U.S. Infantry - Recruiting Detachment: William Bezeau - Age: 23 - Height: 5' 7" - Eyes: Black - Hair: Curly - Complexion: Yellow - Born: Delaware - Trade: Sailor - Enlistment date: 27 Nov 1814 - Enlistment place: Philadelphia - Enlistment period: War - Enlisted by whom: William Bezeau - Comments: Listed as "Col'd" in his service record; deserted on 10 Dec 1814.

White, James - Recruit - 26th U.S. Infantry - Recruiting Detachment: William Bezeau - Age: 23 - Height: 5' 10" - Eyes: Black - Hair: Black - Complexion: Black - Born: Pennsylvania - Trade: Farmer - Enlistment date: 25 Oct 1814 - Enlistment place: Philadelphia - Enlistment period: 5 Yrs - Enlisted by whom: William Bezeau - Discharge date: 17 May 1815 - Comments: Listed as "Col'd" in his service record; discharged at Philadelphia, PA.

White, Thomas - Recruit - 26th U.S. Infantry - Recruiting Detachment: William Bezeau - Age: 27 - Height: 6' - Eyes: Black - Hair: Curly - Complexion: Yellow - Born: Pennsylvania - Trade: Farmer - Enlistment date: 18 Jan 1815 - Enlistment place: Philadelphia - Enlistment period: War - Enlisted by whom: William Bezeau - Comments: Listed on the Descriptive Roll of Colored Men, 1 Apr 1815.

White, William - Recruit - 26th U.S. Infantry - Recruiting Detachment: William Bezeau - Age: 17 -

Height: 4' 8" - Eyes: Black - Hair: Curly - Complexion: Black - Born: Maryland - Trade: Laborer - Enlistment date: 11 Oct 1814 - Enlistment place: Philadelphia - Enlistment period: War - Enlisted by whom: William Bezeau - Comments: Listed as "Colored" and "Col'd" in his service record.

Wigham, Robert - Recruit - 26th U.S. Infantry - Recruiting Detachment: William Bezeau - Age: 28 - Height: 5' 9" - Eyes: Black - Hair: Curly - Complexion: Black - Born: Delaware - Trade: Farmer - Enlistment date: 26 Jan 1815 - Enlistment place: Philadelphia - Enlistment period: War - Enlisted by whom: William Bezeau - Comments: Listed on the Descriptive Roll of Colored Men, 1 Apr 1815.

Willett, Edward (Willitz) - Recruit - 26th U.S. Infantry - Recruiting Detachment: William Bezeau - Age: 21 - Height: 5' 8 1/2" - Eyes: Black - Hair: Black - Complexion: Dark - Born: New Jersey - Trade: Farmer - Enlistment date: 13 Sep 1814 - Enlistment place: Philadelphia - Enlistment period: 5 Yrs - Enlisted by whom: William Bezeau - Discharge date: 20 May 1815 - Comments: Listed as a Negro in his service record; discharged at Philadelphia, PA.

Williams, Cato - Private - 11th U.S. Infantry - Company: John Bliss - Age: 30 - Height: 5' 5 1/2" - Eyes: Black - Hair: Black - Complexion: Black - Born: MA, Berkshire County, Lanesborough - Trade: Farmer - Enlistment date: 24 May 1812 - Enlistment period: 5 Yrs - Enlisted by whom: Samuel Holley - Comments: Possible Black, not verified; absent with the U.S. Navy on Lake Champlain.

Williams, Charles - Private - U.S. Light Artillery - Company: George Morris - Age: 24 - Height: 5' 9" - Eyes: Black - Hair: Black - Complexion: Black or Dark - Born: PA, Philadelphia - Trade: Mariner - Enlistment date: 29 Sep 1814 - Enlistment place: MA, Boston - Enlistment period: War - Enlisted by whom: William Campbell - Discharge date: 31 Mar 1815 - Comments: Possible Black, not verified.

Williams, Henry - Private - U.S. Light Artillery - Company: George Morris - Age: 33 - Height: 5' 5 1/2" - Eyes: Black - Hair: Black - Complexion: Black - Born: Danish West Indies, St. Croix - Trade: Laborer - Enlistment date: 3 Oct 1814 - Enlistment period: War - Enlisted by whom: William Campbell - Comments: Possible Black, not verified.

Williams, John - Recruit - 26th U.S. Infantry - Recruiting Detachment: William Bezeau - Age: 20 - Height: 5' 8" - Eyes: Black - Hair: Curly - Complexion: Yellow - Born: PA, Philadelphia - Trade: Laborer - Enlistment date: 16 Oct 1814 - Enlistment place: Philadelphia - Enlistment period: War - Enlisted by whom: William Bezeau - Discharge date: 23 Mar 1815 - Bounty: BLW 7211-160-12 (Illinois) - Comments: Listed as "Col'd" in his service record; discharged at Philadelphia, PA.

Williams, William - Corporal - 38th U.S. Infantry - Company: Shepperd Leakin - Age: 22 - Height: 5' 8" - Eyes: Grey - Hair: Black - Complexion: Dark - Born: MD, Baltimore - Trade: Laborer - Enlistment date: 5 Apr 1814 - Enlistment period: War - Enlisted by whom: John Martin - Comments: A Mulatto who also used the name "Frederick Hall".

Willis, Henry - Private - 29th U.S. Infantry - Company: John Rochester - Age: 21 - Height: 5' 10" - Eyes: Black - Hair: Black - Complexion: Black - Born: MA, Berkshire County, Stockbridge - Trade: Laborer - Enlistment date: 6 Jul 1814 - Enlistment place: NY, Troy - Enlistment period: 5 Yrs - Enlisted by whom: Gad. Dumbleton - Comments: Possible Black, not verified; deserted at Half Moon, NY, on 25 or 26 Apr 1816.

Wilson, Charles - Recruit - 26th U.S. Infantry - Recruiting Detachment: William Bezeau - Age: 30 - Height: 5' 3" - Eyes: Black - Hair: Black - Complexion: Dark - Born: Africa - Trade: Laborer - Enlistment date: 19 Oct 1814 - Enlistment place: Philadelphia - Enlistment period: War - Enlisted by whom: William Bezeau - Discharge date: 20 Mar 1815 - Bounty: BLW 14827-160-12 (Illinois) - Comments: Listed as "Col'd" in his service record; discharged at Philadelphia, PA.

Wilson, Robert - Recruit - 26th U.S. Infantry - Recruiting Detachment: William Bezeau - Age: 21 - Height: 5' 6" - Eyes: Black - Hair: Curly - Complexion: Yellow - Born: Connecticut - Trade: Whisp maker - Enlistment date: 24 Nov 1814 - Enlistment place: Philadelphia - Enlistment period: 5 Yrs - Enlisted by whom: William Bezeau - Discharge date: 1 Jun 1815 - Comments: Listed as "Colored" in his service record; discharged at Philadelphia, PA.

Wilton – George Welton

Wiman, Ebenezer - Laborer - Corps of Artificers - Company: Alexander Paris - Comments: Listed as "Black" in *A List of Pensioners of the War of 1812*.

Winn, Henry - Recruit - 26th U.S. Infantry - Recruiting Detachment: William Bezeau - Age: 25 - Height: 5' 5" - Eyes: Black - Hair: Black - Complexion: Yellow - Born: Maryland - Trade: Farmer - Enlistment date: 9 Oct 1814 - Enlistment place: Philadelphia - Enlistment period: War - Enlisted by whom: William Bezeau - Discharge date: 20 Mar 1815 - Comments: Listed as "Col'd" in his service record.

Winn, James - Recruit - 26th U.S. Infantry - Recruiting Detachment: William Bezeau - Age: 22 - Height: 5' 5" - Eyes: Black - Hair: Curly - Complexion: Yellow or Black - Born: Maryland - Trade: Laborer - Enlistment date: 10 Oct 1814 - Enlistment place: Philadelphia - Enlistment period: 5 Yrs - Enlisted by whom: William Bezeau - Discharge date: 25 Mar 1815 - Bounty: BLW 10299-160-12 (Illinois) - Comments: Listed as "Col'd" in his service record; discharged at Philadelphia, PA.

Winters, Abraham - Recruit - 26th U.S. Infantry - Recruiting Detachment: William Bezeau - Age: 22 - Height: 5' 6" - Eyes: Black - Hair: Curly - Complexion: Brown - Born: Pennsylvania - Trade: Farmer - Enlistment date: 8 Jan 1815 - Enlistment place: Philadelphia - Enlistment period: War - Enlisted by whom: William Bezeau - Comments: Listed on the Descriptive Roll of Colored Men, 1 Apr 1815.

Winters, Peter - Recruit - 26th U.S. Infantry - Recruiting Detachment: William Bezeau - Age: 26 - Height: 5' 5" - Eyes: Black - Hair: Curly - Complexion: Dark - Born: Delaware - Trade: Seaman - Enlistment date: 7 Dec 1814 - Enlistment place: Philadelphia - Enlistment period: 5 Yrs - Enlisted by whom: William Bezeau - Discharge date: 1 May 1815 - Comments: Listed as "Colored" and as a Negro in his service record; discharged at Philadelphia, PA.

Wright, Samuel W. - Recruit - 26th U.S. Infantry - Recruiting Detachment: William Bezeau - Age: 23 - Height: 5' 6" - Eyes: Black - Hair: Curly - Complexion: Yellow - Born: Virginia - Enlistment date: 14 Jan 1815 - Enlistment place: Philadelphia - Enlistment period: War - Enlisted by whom: William Bezeau - Comments: Listed on the Descriptive Roll of Colored Men, 1 Apr 1815.

Yalpin, Solomon - Recruit - 26th U.S. Infantry - Recruiting Detachment: William Bezeau - Age: 25 - Height: 5' 4" - Eyes: Black - Hair: Curly - Complexion: Black - Born: Pennsylvania - Trade: Porter - Enlistment date: 6 Feb 1815 - Enlistment place: Philadelphia - Enlistment period: 5 Yrs - Enlisted by whom: William Bezeau - Comments: Listed on the Descriptive Roll of Colored Men, 1 Apr 1815.

Yanton, Smith - Recruit - 26th U.S. Infantry - Recruiting Detachment: William Bezeau - Age: 28 - Height: 5' 8" - Eyes: Black - Hair: Curly - Complexion: Black - Born: Pennsylvania - Trade:

Laborer - Enlistment date: 6 Feb 1815 - Enlistment place: Philadelphia - Enlistment period: War - Enlisted by whom: William Bezeau - Comments: Listed on the Descriptive Roll of Colored Men, 1 Apr 1815.

York, Tobias - Recruit - 26th U.S. Infantry - Recruiting Detachment: William Bezeau - Age: 30 - Height: 5' 10" - Eyes: Black - Hair: Black - Complexion: Black - Born: Maryland - Trade: Laborer - Enlistment date: 13 Oct 1814 - Enlistment place: Philadelphia - Enlistment period: War - Enlisted by whom: William Bezeau - Comments: Listed as "Col'd" in his service record; deserted on 11 Nov 1814.

Youell, Daniel Thomas - Private - 30th U.S. Infantry - Company: William Miller - Age: 24 - Height: 5' 8" - Eyes: Black - Hair: Curly - Complexion: Black - Born: NY, New York - Trade: Farmer - Enlistment date: 21 Mar 1814 - Enlistment place: VT, Burlington - Enlistment period: War - Enlisted by whom: William Barney - Comments: Possible Black, not verified; deserted at Plattsburg, NY, on 7 or 11 Sep 1814.

Young, Francis - Recruit - 26th U.S. Infantry - Recruiting Detachment: William Bezeau - Age: 23 - Height: 5' 4" - Eyes: Black - Hair: Curly - Complexion: Black - Born: Pennsylvania - Trade: Brass pounder - Enlistment date: 27 Oct 1814 - Enlistment place: Philadelphia - Enlistment period: War - Enlisted by whom: William Bezeau - Comments: Listed as "Col'd" in his service record; deserted on 11 Nov 1814.

Young, Samuel - Recruit – 26th U.S. Infantry - Recruiting Detachment: William Bezeau - Age: 20 - Height: 5' 9" - Eyes: Black - Hair: Curly - Complexion: Black - Born: Pennsylvania - Trade: Carter - Enlistment date: 28 Jan 1815 - Enlistment place: Philadelphia - Enlistment period: War - Enlisted by whom: William Bezeau - Comments: Listed on the Descriptive Roll of Colored Men, 1 Apr 1815.

Youngsted, Hope (Yousted) - Recruit - 26th U.S. Infantry - Recruiting Detachment: William Bezeau - Age: 24 - Height: 5' 11" - Eyes: Black - Hair: Curly - Complexion: Black - Born: Pennsylvania - Trade: Laborer - Enlistment date: 27 Jan 1815 - Enlistment place: Philadelphia - Enlistment period: War - Enlisted by whom: William Bezeau - Bounty: BLW 811-320-14 Cancelled; BLW 1037-320-14 (Arkansas) - Comments: Listed on the Descriptive Roll of Colored Men, 1 Apr 1815; first bounty was cancelled by an order of the War Department, 19 May 1820.

Notes

Other possible Black Soldiers

Anderson, Clement - Recruit - 26th U.S. Infantry - Company: William Bezeau - Age: 37 - Height: 5' 6" - Eyes: Black - Hair: Black - Complexion: Dark - Born: New Jersey - Trade: Seaman - Enlistment date: 1 Jan 1815 - Enlistment place: Philadelphia - Enlistment period: War - Enlisted by whom: William Bezeau - Comments: Listed on the Descriptive Roll, 1 Apr 1815.

Anthony, James - Recruit - 26th U.S. Infantry - Company: William Bezeau - Age: 41 - Height: 5' 9" - Eyes: Black - Hair: Black - Complexion: Dark - Born: Ohio - Trade: Farmer - Enlistment date: 5 Feb 1815 - Enlistment place: Philadelphia - Enlistment period: War - Enlisted by whom: William Bezeau - Comments: Listed on the Descriptive Roll, 1 Apr 1815.

Bloomer, Samuel - Private - 1st U.S. Infantry - Company: Horatio Stark - Age: 31 - Height: 5' 9" - Eyes: Black - Hair: Brown - Complexion: Brown - Born: New York - Trade: Farmer - Enlistment date: 23 Jun 1812 - Enlisted by whom: John Meeks - Discharge date: 31 May 1814.

Bolton, George Jr - Private - 34th U.S. Infantry - Company: Isaac Carter - Age: 27 - Height: 6' 3/4" - Eyes: Grey - Hair: Black - Complexion: Black - Born: ME, Kennebec County, Augusta - Trade: Farmer - Enlistment date: 25 Mar 1814 - Enlistment place: ME, Portland - Enlistment period: War - Enlisted by whom: William Springer - Comments: Deserted 8 Sep 1814 at Plattsburgh, NY.

Bonner, Joseph - Private - 16th U.S. Infantry - Company: Thomas McMahon - Other regiment: 2nd Infantry - Age: 27 - Height: 5' 2 1/2" - Eyes: Black - Hair: Black - Complexion: Dark or Black - Born: France, Lavautte - Trade: Paver - Enlistment date: 27 Sep 1813 - Enlistment place: PA, Little York - Enlistment period: 5 Yrs - Enlisted by whom: Samuel Wigley - Discharge date: 27 Sep 1818 - Pension: Pension - Comments: Discharged at Sackets Harbor, NY.

Bowen, Thomas - Private - 2nd U.S. Artillery - Company: Nathan Towson - Age: 23 - Height: 5' 5 1/2" - Eyes: Grey - Hair: Brown - Complexion: Brown - Born: Pennsylvania - Trade: Stone mason - Enlistment date: 18 May 1812 - Enlistment place: MD, Baltimore - Enlistment period: 5 Yrs - Enlisted by whom: Nathan Townson - Discharge date: 21 May 1817 - Pension: Pension - Comments: Prisoner of War (Halifax); discharged at Fort Niagara, NY.

Braho, Joseph - Private - 8th U.S. Infantry - Company: Charles Crawford - Age: 22 - Height: 5' 6" - Eyes: Black - Hair: Black - Complexion: Black - Born: Mexico - Trade: Farmer - Enlistment date: 19 Jan 1814 - Enlistment place: GA, Point Peter - Enlistment period: 5 Yrs - Enlisted by whom: James Black - Comments: Deserted on 22 Feb 1815.

Burd, John - Private - 26th U.S. Infantry - Company: William Bezeau - Enlistment date: 26 Oct 1814 - Enlistment place: PA, Philadelphia - Enlistment period: 5 Yrs - Discharge date: 18 May 1815 - Comments: Discharged at Philadelphia, PA.

Burr, William - Private - 37th U.S. Infantry - Company: Stephen Tilden - Age: 35 - Height: 5' 4" - Eyes: Grey - Hair: Black - Complexion: Black - Born: NJ, Sussex County, Greenwich - Trade: Farmer - Enlistment date: 13 May 1814 - Enlistment place: CT, Hartford - Enlistment period: War - Enlisted by whom: Daniel Bicknell - Discharge date: 10 May 1815 - Pension: Pension - Comments: Discharged at New London, CT.

Butt, Thomas - Private - 10th U.S. Infantry - Company: Emanuel Leight - Age: 38 - Height: 5' 8" - Eyes: Grey - Hair: Black - Complexion: Black - Born: NC, Orange County - Trade: Carpenter -

Enlistment date: 11 May 1814 - Enlistment place: NC, Hillsborough - Enlistment period: 5 Yrs - Enlisted by whom: (-----) Bennett - Discharge date: 8 Sep 1815 - Bounty: BLW 26820-160-12 (Arkansas) - Comments: Discharged at Washington, DC; unfit for service.

Champaigne, Lambert (Champygne) - Private - 19th U.S. Infantry - Company: Joel Collins - Other regiment: 3rd U.S. Infantry - Age: 18 - Height: 5' 6" - Eyes: Black - Hair: Black - Complexion: Black - Born: MI, Wayne County, Detroit - Trade: Laborer - Enlistment date: 6 May 1814 - Enlistment place: MI, Wayne County, Detroit - Enlistment period: 5 Yrs - Enlisted by whom: John Meldrum - Comments: Re-enlisted on 18 Jan 1819.

Cotton, Burwell - Private - 2nd U.S. Artillery - Company: John Goodall - Age: 36 - Height: 5' 8" - Eyes: Dark - Hair: Dark - Complexion: Brown - Trade: Cabinet maker - Enlistment date: 23 Jan 1814 - Enlistment place: VA, Abington - Enlistment period: 5 Yrs - Enlisted by whom: James McIntosh - Comments: Deserted from Fort Norfolk, VA, on 22 Mar 1817.

Crane, Asaph - Private - 2nd U.S. Artillery - Company: William Gates - Other regiment: U.S. Light Artillery then Corps of Artillery - Age: 23 - Height: 5' 11" - Eyes: Brown - Hair: Brown - Complexion: Brown - Born: MA, Norfolk County, Canton - Trade: Laborer - Enlistment date: 7 Apr 1812 - Enlistment place: MA, Boston - Enlistment period: 5 Yrs - Enlisted by whom: Thomas Freeman - Discharge date: 6 Apr 1817 - Comments: Re-enlisted.

Dale, Charles - Private - 28th U.S. Infantry - Company: Henry Gist - Other regiment: 5th Infantry - Age: 31 - Height: 5' 9" - Eyes: Black - Hair: Black - Complexion: Yellow or Sallow - Born: Virginia - Trade: Shoemaker - Enlistment date: 9 Sep 1814 - Enlistment place: KY or Augusta - Enlistment period: 5 Yrs - Enlisted by whom: Peter Davis - Comments: Deserted at Detroit, MI, on 26 April 1816.

Davis, David - Private - 2nd U.S. Infantry - Company: Hugh McCall - Other regiment: 1st Infantry then 2nd Infantry - Age: 29 - Height: 6' 2" - Eyes: Black - Hair: Dark - Complexion: Black - Born: Virginia - Trade: Farmer - Enlistment date: 11 Jan 1803 - Enlistment period: 5 Yrs - Died: 5 May 1815 - Comments: Re-enlisted for 5 years on 10 Jul 1813; died at New Orleans, LA.

Davis, John - Private - 24th U.S. Infantry - Company: James Campbell - Age: 22 - Height: 5' 10" - Eyes: Black - Hair: Black - Complexion: Yellow - Born: NC, Nashville - Trade: Farmer - Enlistment date: 14 Jun 1814 - Enlistment place: Tennessee - Enlistment period: War - Enlisted by whom: Abraham Britton - Discharge date: 22 Mar 1815 - Comments: Discharged from Camp Mandeville, LA.

Davis, John - Corporal - 23rd U.S. Infantry - Company: Richard Goodell - Other regiment: 2nd Infantry - Age: 18 - Height: 5' 6" - Eyes: Black or Dark - Hair: Black or Dark - Complexion: Dark - Born: NY, Essex County, Schroon - Trade: Farmer - Enlistment date: 1 May 1814 - Enlistment place: NY, Albany - Enlistment period: 5 Yrs - Enlisted by whom: William Clarke.

Delany, Mack - Private - 10th U.S. Infantry - Company: George Vashon - Other regiment: 8th Infantry - Age: 23 - Height: 5' 10 1/2" - Eyes: Grey - Hair: Dark - Complexion: Yellow - Born: VA, Amelia County - Trade: Blacksmith - Enlistment date: 18 Nov 1812 - Enlistment place: VA, Halifax County - Enlistment period: 5 Yrs - Enlisted by whom: Emanuel Leight - Discharge date: 18 Nov 1817.

Durham, Seth - Private - 8th U.S. Infantry - Company: William Jones - Other regiment: 7th Infantry - Age: 17 - Height: 5' 6" - Eyes: Grey - Hair: Dark - Complexion: Yellow - Born: SC, Barnwell -

Trade: Farmer - Enlistment date: 10 Dec 1813 - Enlistment place: Georgia - Enlistment period: 5 Yrs - Enlisted by whom: (-----) Davis - Comments: Deserted at Fort Hawkins, GA, on 1 Apr 1816.

Eyrick, Jacob - Private - 5th U.S. Infantry - Company: John Taylor - Age: 37 - Height: 5' 1 1/2" - Eyes: Grey - Hair: Brown - Complexion: Yellow - Born: PA, York County - Trade: Weaver - Enlistment date: 13 Aug 1814 - Enlistment place: PA, York - Enlistment period: 5 Yrs - Enlisted by whom: John Fendall.

Forman, Aaron - Private - 2nd U.S. Infantry - Company: Mathew Arbuckle - Age: 20 - Height: 5' 7" - Eyes: Grey - Hair: Black - Complexion: Yellow - Born: VA, Berkley County - Trade: Farmer - Enlistment date: 13 Jun 1808 - Enlistment place: MD, Fredericktown - Enlistment period: 5 Yrs - Enlisted by whom: John Luckett - Died: 11 Dec 1813 - Comments: Died at Fort Charlotte, AL.

Frey, Michael - Private - 32nd U.S. Infantry - Company: George Goodman - Age: 26 - Height: 5' 6 3/4" - Eyes: Black - Hair: Dark - Complexion: Black - Born: Pennsylvania - Trade: Farmer - Enlistment date: 17 Feb 1814 - Enlistment place: PA, Easton - Enlistment period: War - Enlisted by whom: Henry Gulick - Discharge date: 7 May 1815 - Bounty: BLW 4407-160-12 (Illinois) - Comments: Discharged at Governor's Island, NY.

Graves, George B. - Private - 14th U.S. Infantry - Company: Joseph Marshall - Age: 27 - Height: 5' 11 1/2" - Eyes: Dark - Hair: Black - Complexion: Dark - Born: VA, Portsmouth - Trade: Tailor - Enlistment date: 2 Aug 1814 - Enlistment place: VA, Alexandria - Enlistment period: War - Enlisted by whom: James Riddle - Comments: Also served as a nurse in the general hospital at Greenbush, NY.

Hagar, Francis - Private - 9th U.S. Infantry - Company: William Foster - Age: 40 - Height: 5' 9" - Eyes: Black or Light - Hair: Black - Complexion: Black - Born: MA, Essex County, Watertown - Trade: Farmer - Enlistment date: 28 Jan 1813 - Enlistment place: MA, Watertown - Enlistment period: 5 Yrs - Enlisted by whom: Ebenezer Thompson - Discharge date: 3 Jul 1815 - Pension: Pension - Comments: Discharged at Sackets Harbor, NY, because of severe wounds.

Hagins, Jonathan (Hagans) - Private - 10th U.S. Infantry - Company: Robert Mitchell - Other regiment: 8th Infantry - Age: 22 - Height: 5' 9" - Eyes: Dark - Hair: Black - Complexion: Yellow - Born: NC, Sampson County - Trade: Farmer - Enlistment date: 14 Jul 1813 - Enlistment place: NC, Charlotte - Enlistment period: 5 Yrs - Enlisted by whom: Alexander Brandon - Discharge date: 23 May 1818 - Comments: Discharged at Fort Crawford, IL.

Haskell, James - Private - U.S. Light Artillery - Company: Benjamin Branch - Age: 18 - Height: 5' 10" - Eyes: Black - Hair: Dark or Black - Complexion: Black or Dark - Born: MA, Cumberland County, Standish - Trade: Farmer - Enlistment date: 23 Feb 1813 - Enlistment place: ME, Fort Preble - Enlistment period: War - Enlisted by whom: Thomas Pitts - Discharge date: 31 May 1815 - Comments: Discharged at Plattsburg, NY.

Hines, William - Private - 38th U.S. Infantry - Company: James Smith - Age: 21 - Height: 5' 7" - Eyes: Brown - Hair: Dark - Complexion: Black - Born: MD, Hartford County - Trade: Seaman - Enlistment date: 20 Feb 1814 - Enlistment place: VA, Craney Island - Enlistment period: War - Enlisted by whom: James Haslett - Discharge date: 15 Mar 1815 - Pension: Pension - Comments: Discharged at Craney Island, VA.

Hoggard, James (Hogart, Hoggart) - Private - 10th U.S. Infantry - Company: George Vashan - Age: 20 - Height: 5' 6" - Eyes: Black - Hair: Black - Complexion: Yellow - Born: NC, Bertie County -

Trade: Farmer - Enlistment date: 3 Jul 1812 - Enlistment period: 5 Yrs - Enlisted by whom: Richard Plummer - Discharge date: 3 Aug 1817 - Pension: Pension - Comments: Discharged at Camp Experiment, AL.

Irick – see Jacob Eyrick

Johnson, James - Private - 14th U.S. Infantry - Company: David Cummings - Age: 19 - Height: 5' 5" - Eyes: Dark - Hair: Brown - Complexion: Yellow - Born: MD, Baltimore - Trade: Shoemaker - Enlistment date: 6 Nov 1814 - Enlistment place: Camp Snowden - Enlistment period: 5 Yrs - Enlisted by whom: James Riddle.

Jones, Rowland - Fifer - 10th U.S. Infantry - Company: Robert Mitchell - Other regiment: 8th Infantry - Age: 28 - Height: 5' 9" - Eyes: Black - Hair: Black - Complexion: Yellow - Born: VA, Lunenburg - Trade: Ditcher - Enlistment date: 19 Jun 1812 - Enlistment place: NC, Person - Enlistment period: 5 Yrs - Enlisted by whom: Robert Mitchell - Discharge date: 19 Jun 1817 - Bounty: BLW 26801-160-12 (Arkansas) - Comments: Discharged at Pass Christian, MS.

McCoy, Archibald - Private - U.S. Rifles - Company: S V Hamilton - Age: 25 - Height: 5' 8" - Eyes: Brown - Hair: Brown - Complexion: Brown - Born: NY, Sussex County - Trade: Farmer - Enlistment date: 24 Apr 1814 - Enlistment place: PA, Easton - Enlistment period: War - Enlisted by whom: Capt Morgan - Discharge date: 24 Aug 1815 - Comments: Discharged at Plattsburgh, NY.

Merry, Joseph (Meny) - Private - U.S. Light Artillery - Company: John Bell - Age: 22 - Height: 5' 4" - Eyes: Black or Dark - Hair: Black or Dark - Complexion: Dark - Born: Spain, Havana (Cuba) - Trade: Mariner - Enlistment date: 20 Jun 1814 - Enlistment place: MA, Boston - Enlistment period: War - Enlisted by whom: Samuel Washburn - Discharge date: 31 Mar 1815.

Montgomery, John - Private - 13th U.S. Infantry - Company: Mordecai Myers - Age: 34 - Height: 5' 8" - Eyes: Dark - Hair: Dark - Complexion: Black - Born: VT, New Derry - Trade: Farmer - Enlistment date: 29 Sep 1814 - Enlistment place: Auburn - Enlistment period: 5 Yrs - Enlisted by whom: Henry Minton.

Nelson, John - Private - 2nd U.S. Artillery - Company: Philemon Hawkins - Age: 22 - Height: 6' 2" - Eyes: Black - Hair: Dark - Complexion: Dark - Born: West Indies - Trade: Carpenter - Enlistment date: 11 Dec 1814 - Enlistment period: War - Enlisted by whom: Lt Guy - Comments: Discharged at end of war.

Nelson, Robert H. - Private - U.S. Light Artillery - Company: Samuel Spotts - Age: 31 - Height: 6' - Eyes: Grey - Hair: Brown - Complexion: Black - Born: New York - Trade: Farmer - Enlistment date: 22 Mar 1813 - Enlistment place: LA, New Orleans - Enlistment period: War - Enlisted by whom: Samuel Spotts - Discharge date: 9 Apr 1815.

Orteza, Dominic - Musician - 5th U.S. Infantry - Company: James Dorman - Height: 5' 4" - Eyes: Black - Hair: Black - Complexion: Black - Born: Italy - Trade: Musician - Enlistment date: 1 May 1811 - Enlistment place: DC, Georgetown - Enlistment period: 5 Yrs - Enlisted by whom: J Johnson.

Parcels, Isaac - Private - U.S. Light Artillery - Company: Jonathan Brooks - Age: 25 - Height: 5' 10" - Eyes: Black - Hair: Black - Complexion: Black - Born: NY, South or Turtle Bay - Trade: Carpenter - Enlistment date: 4 Dec 1812 - Enlistment place: New York - Enlistment period: 5 Yrs - Enlisted by whom: Jonathan Brooks - Discharge date: 4 Dec 1817.

Parrish, John - Private - 8th U.S. Infantry - Company: William Chisolm - Other regiment: 7th Infantry - Age: 18 - Height: 5' 8 1/2" - Eyes: Black - Hair: Dark - Complexion: Yellow - Born: GA, Columbia - Trade: Farmer - Enlistment date: 6 Oct 1812 - Enlistment place: Watkinsville - Enlistment period: 5 Yrs - Enlisted by whom: John Mallory - Discharge date: 6 Oct 1817 - Pension: Pension - Comments: Discharged at Fort Crawford, IL.

Porter, John - Private - 21st U.S. Infantry - Company: Morrill Marston - Age: 39 - Height: 5' 6 1/2" - Eyes: Hazel - Hair: Black - Complexion: Black - Born: Germany - Trade: Mariner - Enlistment date: 30 May 1814 - Enlistment place: NH, Portsmouth - Enlistment period: 5 Yrs - Enlisted by whom: John Dix - Died: 20 Apr 1815 - Comments: Died in Brownsville, NY.

Roberts, John - Private - 8th U.S. Infantry - Company: John Allison - Other regiment: 7th Infantry - Age: 26 - Height: 5' 10 1/2" - Eyes: Black - Hair: Black - Complexion: Black - Born: VA, Charlotte - Trade: Carpenter - Enlistment date: 4 Oct 1814 - Enlistment place: GA, Fort Hawkins - Enlistment period: 5 Yrs - Enlisted by whom: Lieutenant Womack - Comments: Deserted at Fort Hawkins, GA, on 1 May 1816.

Roberts, Wiley - Private - 18th U.S. Infantry - Company: Benjamin Elmore - Other regiment: 4th Infantry - Age: 18 - Height: 5' 7" - Eyes: Black or Blue - Hair: Black or Dark - Complexion: Black or Fair - Born: SC, Laurens - Trade: Farmer - Enlistment date: 21 Apr 1814 - Enlistment place: Newburg - Enlistment period: 5 Yrs - Enlisted by whom: George Bruton - Discharge date: 1 Apr 1819.

Scaudling Jacob - Private - 38th U.S. Infantry - Company: John Rotbrock - Age: 27 - Height: 5' 11" - Eyes: Dark - Hair: Black - Complexion: Black - Born: PA, Chester County - Trade: Farmer - Enlistment date: 10 Aug 1813 - Enlistment place: VA, Craney Island - Enlistment period: War - Enlisted by whom: John Rotbrock - Discharge date: 15 Mar 1815 - Pension: Pension - Comments: Discharged at Craney Island, VA.

Scriver, John - Private - 23rd U.S. Infantry - Company: Horatio Armstrong - Age: 21 - Height: 5' 3" - Eyes: Black - Hair: Black - Complexion: Black - Born: KY, Kingston County, Sopus - Trade: Blacksmith - Enlistment date: 4 Feb 1813 - Enlistment place: NY, Duchess County, Rhinebeck - Enlistment period: 5 Yrs - Enlisted by whom: Horatio Armstrong - Discharge date: 5 Nov 1817 - Comments: Re-enlisted after discharge.

Sharp, Solomon (Sharpe) - Private - 11th U.S. Infantry - Company: Benjamin Smead - Other regiment: 6th U.S. Infantry - Age: 29 - Height: 5' 10" - Eyes: Black - Hair: Black - Complexion: Black - Born: MA, Conway - Trade: Farmer - Enlistment date: 2 Feb 1812 - Enlistment place: VT, Bennington - Enlistment period: 5 Yrs - Enlisted by whom: Benjamin Smead - Discharge date: 2 Feb 1818 - Bounty: BLW 16170-160-42 (Illinois) - Comments: Prisoner of War captured on 3 Jun 1813 on the U.S. Sloop Growler on Lake Champlain, exchanged at Chazy, 11 May 1814.

Springsteen, John - Private - 41st U.S. Infantry - Company: Mangle Quackenbos - Age: 29 - Height: 5' 9" - Eyes: Black - Hair: Dark - Complexion: Black - Born: NY, Dutchess County, Poughkeepsie - Trade: Shoemaker - Enlistment date: 25 Aug 1814 - Enlistment place: NY, Hudson - Enlistment period: War - Enlisted by whom: Theophilus Beekman - Comments: Deserted on 24 Apr 1815.

Sutton, Dempsey - Private - 24th U.S. Infantry - Company: James Campbell - Other regiment: 7th Infantry - Age: 26 - Height: 5' 10" - Eyes: Black - Hair: Black - Complexion: Yellow - Born: North Carolina - Trade: Farmer - Enlistment date: 13 Dec 1813 - Enlistment place: TN, Wilson County -

Enlistment period: 5 Yrs - Enlisted by whom: (-----) Hays - Discharge date: 13 Dec 1818 - Comments: Discharged at Fort Gadsden, FL.

Thomas, Jeremiah - Sergeant - 16th U.S. Infantry - Company: W M Oliver - Age: 23 - Height: 5' 8" - Eyes: Brown - Hair: Black - Complexion: Yellow - Born: Pennsylvania - Enlistment date: 23 Oct 1813 - Enlistment place: PA, Carlisle - Enlistment period: War - Enlisted by whom: ENS Downey - Discharge date: 1 Jun 1815 - Comments: Discharged at Buffalo, NY.

Warman, Thomas - Recruit - 26th U.S. Infantry - Company: William Bezeau - Age: 36 - Height: 6' - Eyes: Dark - Hair: Dark - Complexion: Brown - Born: New Jersey - Trade: Cooper - Enlistment date: 16 Jan 1815 - Enlistment place: Philadelphia - Enlistment period: War - Enlisted by whom: William Bezeau - Comments: Listed on the Descriptive Roll of 1 Apr 1815; deserted on 24 Jan 1815.

Watson, Allen - Private - 26th U.S. Infantry - Company: William Bezeau - Age: 21 - Height: 5' 2" - Eyes: Dark - Hair: Brown - Complexion: Brown - Born: Pennsylvania - Trade: Farmer - Enlistment date: 18 Jan 1815 - Enlistment place: Philadelphia - Enlistment period: War - Enlisted by whom: William Bezeau - Comments: Listed on the Descriptive Roll of 1 Apr 1815.

Willis, Henry - Private - 29th U.S. Infantry - Company: John Rochester - Other regiment: 6th Infantry - Age: 21 - Height: 5' 10" - Eyes: Black - Hair: Black - Complexion: Black - Born: MA, Stockbridge - Trade: Laborer - Enlistment date: 6 Jul 1814 - Enlistment place: NY, Troy - Enlistment period: 5 Yrs - Enlisted by whom: Gadius Dumbleton.

Winsell, Philip (Wensell) - Private - 1st U.S. Infantry - Company: Hugh Moore - Age: 25 - Height: 5' 7" - Eyes: Hazel - Hair: Black - Complexion: Black - Born: PA, Westmoreland County - Trade: Farmer - Enlistment date: 24 Feb 1812 - Enlistment place: PA, Greensburg - Enlistment period: 5 Yrs - Enlisted by whom: McGunigle - Died: 25 Feb 1814 - Comments: Died at Fort Wayne, IN, from dropsy.

Youngblood, Joseph - Private - 24th U.S. Infantry - Company: Robert Desha - Age: 26 - Height: 5' 5 1/2" - Eyes: Black - Hair: Black - Complexion: Yellow - Born: SC, Charleston - Trade: Carpenter - Enlistment date: 5 Feb 1815 - Enlistment place: TN, Nashville - Enlistment period: War - Enlisted by whom: Lt Perkins - Discharge date: 26 Apr 1815 - Bounty: BLW 260-320-42 (Illinois) - Comments: Discharged at Nashville, TN.

The Bibliography

Adjutant and Inspector General's Office, **Military Laws and Rules and Regulations for the Army of the United States**, September 1816, (E. De Krafft, Printer: Washington, DC).

Clark, Byron N., **A List of Pensioners of the War of 1812**, (Research Publications Company: Burlington and Boston, 1904), Payroll of a Company of Artificers, Commanded by Alexander Parris, Superintendent of the Corps of Artificers of the United States, for the months of May and June 1814, pages 129-131.

Heitman, Francis B., **Historical Register and Dictionary of the United States Army From Its Organization, September 29, 1789, to March 2, 1903**, Volume I, (Genealogical Publishing Company, Baltimore, Maryland: 1994).

National Archives and Records Administration, **Index to the Compiled Military Service Records for the Volunteer Soldiers Who Served During the War of 1812,** (Washington, D.C.: National Archives and Records Administration), M602, 234 rolls.

National Park Service, U.S. Department of the Interior, **39th Infantry Muster Roll**, May 1814, Battle of Horseshoe Bend, http://www.nps.gov/hobe/upload/39infantry.pdf.

Official Correspondence with the Department of War relative to the Military Operations of the American Army under the Command of Major General Izard of the Northern Frontier of the United States in the Years 1814 and 1815, (William Fry, Printer: Philadelphia, 1816), pp. 45-46, Letter from Major General Izard to the Secretary at War, 3 July 1814.

Peters, Richard, **The Public Statutes at Large of the United States of America**, Volumes I and II, (Charles C. Little and James Brown: Boston, MA 1845).

Records of Officers and Men of New Jersey in Wars 1791-1815, (Adjutant-General's Office: Trenton, NY 1909), part V, War with Great Britain 1812-1815, Twenty-Sixth Regiment, Infantry – United States Army, pp. 217-218.

Records Relating to War of 1812 Prisoners of War, 1812; (Washington, DC: National Archives Microfilm Publication M2019); Records of the Adjutant General's Office, 1780's-1917; Record Group 94, compiled by Clair Prechtel-Kluskens.

Register of Enlistments in the U.S. Army, 1798-1914; (Washington, DC: National Archives Microfilm Publication M233, 81 rolls); Records of the Adjutant General's Office, 1780's-1917, Record Group 94.

Stagg, J.C.A., **Enlisted Men in the United States Army, 1812-1815: A Preliminary Survey**, *William and Mary Quarterly,* Third Series, Volume XLIII, Number 4, October 1986, pages 615-645.

United States. Bureau of Land Management, General Land Office Records. Automated Records Project; **Federal Land Patents**, State Volumes. http://www.glorecords.blm.gov/. Springfield, Virginia: Bureau of Land Management, Eastern States, 2007.

United States Army, 39th Infantry Muster Roll, National Park Service, Horseshoe Bend National Military Park, Alabama, regimental muster rolls completed after the Battle of Horseshoe Bend, 27 March 1814.

War of 1812 Military Bounty Land Warrants, 1815-1858; (Washington, DC: National Archives Microfilm Publication M848, 14 rolls); Records of the Bureau of Land Management, Record Group 49.

War of 1812 Pension Applications; (Washington D.C.: National Archives, NARA Microfilm Publication M313, 102 rolls), Records of the Department of Veterans Affairs, Record Group Number 15.

Made in United States
Troutdale, OR
11/03/2026

56845389R00042